D0140931

Chance or Reality
and Other Essays

By the same author:

Les tendances nouvelles de l'ecclésiologie

The Relevance of Physics

Brain, Mind and Computers
(Lecomte du Noüy Prize, 1970)

The Paradox of Olbers' Paradox

The Milky Way: An Elusive Road for Science

*Science and Creation: From Eternal Cycles
to an Oscillating Universe*

*Planets and Planetarians: A History of Theories of the Origin
of Planetary Systems*

The Road of Science and the Ways to God
(Gifford Lectures, University of Edinburgh, 1975 and 1976)

The Origin of Science and the Science of its Origin
(Fremantle Lectures, Oxford, 1977)

*And on This Rock: The Witness of One Land
and Two Covenants*

Cosmos and Creator

Angels, Apes, and Men

Uneasy Genius: The Life and Work of Pierre Duhem

Chesterton, a Seer of Science

The Keys of the Kingdom: A Tool's Witness to Truth

Lord Gifford and His Lectures: A Centenary Retrospect

————————

Translations with introduction and notes:

The Ash Wednesday Supper (Giordano Bruno)

*Cosmological Letters on the Arrangement
of the World Edifice* (J.-H. Lambert)

Universal Natural History and Theory of the Heavens (I. Kant)

Stanley L. Jaki

Chance or Reality
and Other Essays

University Press of America

The Intercollegiate Studies Institute, Inc.

Copyright © 1986 by

University Press of America,® Inc.

4720 Boston Way
Lanham, MD 20706

3 Henrietta Street
London WC2E 8LU England

All rights reserved

Printed in the United States of America

**Co-published by arrangement with the
Intercollegiate Studies Institute, Inc.**

Library of Congress Cataloging in Publication Data

Jaki, Stanley L.
  Chance or reality and other essays.

  Includes index.
    1. Science—Philosophy.   2. Physics—Philosophy.
  3. Science—Social aspects.   I. Title.
  Q175.3.J35   1986   501   86-19078
  ISBN 0-8191-5656-6 (alk. paper)
  ISBN 0-8191-5657-4 (pbk. : alk. paper)

39726

JUL 2 8 1987

All University Press of America books are produced on acid-free
paper which exceeds the minimum standards set by the National
Historical Publications and Records Commission.

Q
175.3
J35
1986

Contents

Preface

A dozen essays, written well over a dozen years and about topics as different as their titles suggest, may not readily be assumed to have a unitary theme. Their reprinting in a single volume was, however, prompted not so much because some of them appeared in periodicals and publications of modest circulation as because a theme was common to them. That theme is the concern for what are usually referred to as the humanistic values or perspectives which seem to lose their significance owing to the ever more pervasive presence of the sciences in every facet of human life.

This is not to suggest that the cultivation of science is not an eminently human pursuit. But the single-minded attention which in scientific work is given to quantitative correlations cannot help weakening sensitivity for the realm of qualities or values. In these essays that realm is taken not so much in a wide as in a deep or fundamental sense.

The title of the first essay, chance or reality, stands for a basic contention of these essays: qualities and values cannot be safeguarded unless they are anchored in a renewed respect for their ontological ground. Nothing seems to undermine that ground so effectively as that connection which is widely believed to exist, though without any good ground, between a haphazard notion of existence and the basic lessons of the sciences, especially of modern physics. A defense of objective reality—also the purpose of the second essay—is in the best interest of the sciences as well if they are not to be exploited by sheer subjectivism.

Subjectivism is at work whenever science is taken to be equivalent to a philosophy which, for all its claim to be purely scientific, is a mere scientism, or a rank discredit to the sciences as well as to philosophy. The true physiognomy of scientism as a dehumanizing force is presented in the third and fourth essays through the reflections of two great humanists, Maritain and Chesterton, both of whom had deep appreciation for and penetrating insights into the true meaning of the sciences. A third humanist, Goethe, discussed in the fifth essay, is a classic case of the radical misunderstanding about the exact sciences in which a humanist, however great, can be trapped by his own dabbling in them.

The mutual misunderstanding of men of letters and men of science is the topic of the sixth and seventh essays. The former presents the history of two cultures for the past hundred years, the latter contrasts the respective features of scientific and humanistic knowledge. They are followed by two essays on the relation which can be established between the sciences and that deepest level of humanities where man, a religious animal according to Burke's famed dictum, claims his rights. Theology as humanism touches on man's most universal aspirations which in the tenth essay are set in the perspective of the modern university and of the role to be occupied in its curriculum by the universe as the greatest discovery of modern science.

Science is then investigated in the deeply theologico-humanistic matrix of its birth in two essays relating to the fate and fortunes of science among the Greeks of old and within the Christian cultural context. Prominent place is given in both to the epoch-making findings of Pierre Duhem as a historian of science. Those findings alone make it possible to raise the question broached in the last essay about the role of science in the ultimate outcome of human history, in a manner which permits us to go beyond the shallows of historical relativism and pragmatism, where science may turn into the ultimate nemesis of mankind.

Each essay is reprinted here with practically no alteration. Only a modest effort was made at introducing some uniformity into the footnote style which almost each and every editor prefers to make as peculiar as possible. The repeated appearance of some details may help the reader have a glimpse of my preferences and of some of my deepest convictions about their importance. This should be particularly true of my recourses to the drama told by Plato in the *Phaedo*: its impact on the subsequent development of the humanities and the sciences in the Western world cannot be emphasized enough. Appreciation of such perspectives by Mr. Gregory Wolfe, Publications Director of The Intercollegiate Studies Institute and editor of *The Intercollegiate Review,* played no small part in making this volume a reality.

S.L.J.

1

Chance or Reality:
Interaction in Nature
Versus Measurement in Physics

A title like this may seem to be an invitation into the realm of scientific and philosophical abstractions. Flesh and blood reality calls no less for such a title. A telling illustration of this was provided less than half a year ago in a TIME essay which dealt with "the importance of being lucky." The essay was written with an eye on President Reagan's luck. Indeed the chances were astronomical against the stopping of the assassin's bullet at 2 cm from the President's heart. Had the author of that essay waited two

Paper read during the 5th International Humanistic Symposium organised by the "Hellenic Society for Humanistic Studies" in Portaria/Pelion (September 16-22, 1981) on "Freedom and Necessity in European Civilization. Perspectives of Modern Consciousness." Permission to publish the paper immediately in *Philosophia* (Athens) 10-11 (1981), pp. 85-102, was given by the Hellenic Society, which published in full the Proceedings of the Symposium in 1985.

more weeks, he could have referred to another extremely lucky shot. That no vital organs—liver, pancreas, spinal column—were injured as a bullet criss-crossed through the pope's abdomen, could appear the kind of luck which is better called a miracle. Our secular culture, proud of its rationality, prefers to speak of mere luck. But as the TIME essayist pointed out, "people who believe in luck are not particularly rationalist either, since scientific rationalism has as much trouble dealing with luck as theology does. The best it [scientific rationalism] has to offer is Heisenberg's uncertainty principle, which states that it is absolutely impossible to predict the exact behavior of atomic particles."[1]

That an essayist who writes for a weekly magazine like TIME is allowed by the editor to mention Heisenberg's uncertainty principle, is a proof that the average educated man has already heard of it. It would be easy to show that the principle has become, since its formulation in 1927, as much a part of the cultural atmosphere as was the case a hundred years ago with Darwinian natural selection, and with Newtonian physics over two hundred years ago. Both about Heisenberg's principle and of Darwin's natural selection it may also very well be true what Voltaire said a few years after Newton's death: "Very few people read Newton because it is necessary to be learned in order to understand him. Yet everybody talks about him."[2]

Newton's laws imply that the future position of a particle can be predicted if one knows exactly the position (x), the mass (m), and the velocity (v) of a particle at a given moment (t). Heisenberg's principle states that a simultaneous measurement of these parameters cannot be done with lesser uncertainty than Planck's quantum (h). Or as the now famous formula has it, $\Delta x \cdot \Delta mv \geqslant h$. The formula readily lends itself to the transformation into $\Delta E \cdot \Delta t \geqslant h$, which expresses the minimum uncertainty in the simultaneous measurement of energy (E) and time (t). Both formulas have their equivalents for rotational motion. As almost all other major discoveries in physics, these uncertainty relations too had been in the air for some time before they received in Heisenberg's hands a derivation from general theoretical principles in 1927. A combination of the Compton effect and the De Broglie matter-wave formula could have yielded the Heisenberg formula already in 1924. That calculations of atomic processes are probabilistic in character had also been recognized before Max Born showed in 1926 that the ψ

function, or the very core of Schrödinger's wave mechanics, implies a probability distribution. So much in the way of technicalities about the uncertainty principle which in itself expresses the limits set to the precision of measuring physical interactions, limits which become significant only on the atomic level.

That there are limits, at least practical limits, to the precision achievable in physical measurements had been recognized by all physicists for centuries before Heisenberg. Descartes, Galileo, and Newton, all spoke of the difference between the exactness of all physical interactions and the invariable inaccuracy of their being measured by physicists. Theories of error and theories of the distribution of observational data about the most probable value (Gauss' curve) had been worked out in the early 19th century and systematically applied. The same century was not over yet when it became recognized through Lord Rayleigh's work that the wave nature of light sets a limit to the precision of optical instruments. Even this limitation did not appear worrisome as in principle ever shorter wavelengths could be resorted to. In other words, prior to Heisenberg physicists could safely believe in the limitless perfectibility of their measurements and in the agreement of any ideally perfect measurement with reality which obeyed complete exactness. In Laplace's well known statement, a superior intellect to whom all data were available, could calculate and predict all future configurations of all material bodies with complete precision. Everything, added Laplace, followed the paths of exact mechanical causality, not only the planets whose orbits even ordinary intellects could calculate with almost complete precision, but also individual vapor molecules about whose trajectories nothing could be determined.[3]

The recognition that owing to Heisenberg's principle a theoretical limit prevailed over the perfectibility of instruments and observations did not in itself threaten belief in full physical determinism. True, it was no longer possible to hold that the ideal of perfect precision could ever be achieved. Yet in most areas of physics there remained plenty of room for making measurements more accurate and, far more importantly, it could also be argued that the absence of complete precision in measurements was a statement very different from absence of complete determinism, or physical causality, in the interactions themselves. Insofar as the Heisenberg principle was taken as a mere limit on precision, it was still possible to view chance as being opposite to cause and reality. The view which became a vogue during the Age of Reason, had in its grip not only physicists but also philosophers and poets too. "What we call chance is not and cannot be except the unknown cause of a known effect," declared Voltaire.[4] Schiller may have paraphrased Voltaire as he put in Wallenstein's mouth the words: "Happenstance does not exist."[5]

In both those statements, and many others could be quoted, chance is taken in the sense of non-entity, or the opposite of reality. The same sense also turns up in the late 19th century in a very important context, in T. H. Huxley's reminiscences on the reception of Darwin's theory. There Huxley took to task those who rejected Darwinism on the ground that it was a "reign of chance." "Do they believe," Huxley asked, "that anything in this universe happens without reason or without a cause? Do they really conceive that any event has no cause, and could not have been predicted by any one who had sufficient insight into the order of Nature?" The second question evoked Laplace as did Huxley's description of a seashore where apparently nothing could be measured about the trajectory of individual vapor molecules arising from myriads of bubbles and flakes of foam. But, Huxley warned, "the man of science knows that here, as everywhere, perfect order is manifested; that there is not a curve of the waves, not a note in the howling chorus, not a rainbow glint on a bubble which is other than a necessary consequence of the ascertained laws of nature; and that with sufficient knowledge of the conditions competent physico-mathematical skill could account for, and indeed predict, every one of those 'chance' events." A scientist, Huxley declared, is a convert with only one act of faith, which is "the confession of the universality of order and of the absolute validity in all time and under all circumstances of the law of causation."[6]

Today, all Darwinists and almost all evolutionists speak in a manner of which the title of J. Monod's famous book, *Le Hasard et la Nécéssité*, is a capsule formula.[7] They think that chance and necessity can coexist in the very same process because they almost invariably endorse a dismissal of causality which Heisenberg was the first to tack on the principle of uncertainty. Already in 1927 Heisenberg declared: "Since all experiments are subjected to the laws of quantum mechanics and thereby to the equation $[\Delta x \cdot \Delta mv \geqslant h]$, the invalidity of the law of causality is definitely proved by quantum mechanics.'"[8] Clearly, this meaning given to the uncertainty principle should seem very drastic in comparison with the one which merely states the inability of physicists to secure precision to their measurements beyond a certain limit. Most educated laymen (at least in the Anglo-Saxon world) learned about that drastic meaning from Eddington's books and addresses. As early as 1927, he spoke of the emergence in the new physics "of an attitude more definitely hostile to determinism."[9]

The rest is fairly well known, although some highlights are worth recalling. In accepting in 1933 the Nobel Prize, Heisenberg stated that on account of the principle of uncertainty one must forego "the objectification [of knowledge] to a greater extent than hitherto expected."[10] About the same time, John von Neumann analyzed the possibility of constructing a theory of which quantum mechanics with its probability calculations would be a particular case. That broader theory (later spoken of as a theory of hidden variables) would allow, in principle at least, complete precision in measurement and therefore exact predictability of atomic events. The answer was negative and von Neumann saw in this an imperative to endorse the drastic meaning of Heisenberg's principle: "There is at present no reason to speak of causality in nature—because no experiment indicates its presence and . . . quantum mechanics contradicts it." Causality, von Neumann added, was an age-old way of thinking which "has been done away with."[11]

Causality could easily be rejected in a philosophical atmosphere which had grown increasingly skeptical since the days of Hume and Kant. No more telling sign of that atmosphere would perhaps ever be found than the readiness of most leading physicists to accept such consequences of the rejection of causality as the denial of strict interaction at the basic level of nature and the dismissal of objective reality itself. Once statements about causality became a question of strict measurability, processes involving single atoms no longer retained any meaning. The radioactive decay of an atom could only be seen as a succession of states with no connection or interaction among those states.

The characterization of natural radioactivity as a spontaneous disintegration, a patently anthropomorphic term, was much more expressive of the logic at work than could appear on a cursory look. Once objective causality was abandoned, it became almost unavoidable to attribute volition to atoms in order to retain the semblance of coherent discourse and of a coherent nature. This came into the open already in 1927 in connection with cloud chamber tracks. While the visible track was a unity, no connection could be assigned to the millions of ionized molecules because interaction between any two of them was not measurable. Was their succession, resulting in an obvious unity, a choice on the part of Nature? To answer affirmatively this question posed by Dirac to Heisenberg[12] was of course inadmissible within a science which since the days of Galileo and Descartes excluded from its domain any volition, purpose, and goal. But was it not even more incompatible with the spirit of physics to declare, as Heisenberg did, a purely physical situation non-existent because it was uninvestigable in the sense of not being exactly measurable? Indeed, was it in the spirit of the same science to doubt on the same ground the existence of external reality itself?

In this crucially decisive respect too the force of logic operating within any premise, especially if drastic, was not only inexorable but speedy as well. On the basis that it was impossible to differentiate with complete precision in space and time between the radioactive and non-radioactive isotopes of any atom, say potassium (kalium), Eddington declared in 1932: "The answer of modern physics is that strictly speaking there is not such a thing as a K_{39} atom but only an atom which has a high probability of being K_{39}."[13] From mere probability with respect to being it was only a short step to despair about the rationality of existence. Once more the physicists-turned-philosophers obeyed the call of logic with no delay. In speaking to a world-wide educated public on the pages of *Harper's Magazine* about the new vision of science, the future Nobel laureate, P. W. Bridgman, declared: "The world is not a world of reason, understandable by the intellect of man . . . the world is not intrinsically reasonable or understandable." If such was the case the reality of the world itself was in question and Bridgman knew it: "A vision has come to the physicist in this experience which he will never forget; the possibility that the world may fade away, elude him, and become meaningless because of the nature of knowledge itself [that possibility] has never been envisaged before, at least by the physicist, and this possibility must forever keep him humble."[14] Yet if anyone, were he the most prominent of all physicists, discoursed about knowledge on the shallow grounds that all its merits stood and fell with exact measurability, the humility advocated in the same breath would not be any deeper and more persuasive than the discourse itself. Journalists hardly ever mindful of humility or of logic felt themselves free of all constraints in presenting to the general public the new world view imposed by atomic physics. In reminiscing of this in 1950, the prominent physicist H. Margenau wrote: "No simple slogan, save 'violation of causal reasoning' was deemed sufficiently dramatic to describe the revolutionary qualities of the new knowledge."[15]

Prominent voices of opposition were not lacking but they were largely drowned out in the celebration of non-causal reasoning as reality was given innumerable fare-well parties. Today, only some historians of science recall Einstein's famous declaration which he sent to *Nature* as his contribution to the bicentenary of Newton's death in 1927: "It is only in the quantum theory that Newton's differential method becomes inadequate, and indeed strict causality fails us. But the last word has not yet been said. May the spirit of Newton's method give us the power to restore unison between physical reality and the profoundest characteristic of Newton's teaching—strict causality."[16] The statement revealed both Einstein's instinctive attachment to causality and also his articulateness as a philosopher. He failed to see that lack of precision in measurements and predictions is not logically equivalent to absence of causality. Much lesser minds than Einstein perceived clearly, right there and then, this all important point. Nothing in the way of rigor can indeed be added to the concluding phrase of a letter written in 1930 to the editor of *Nature* by a today completely forgotten teacher at the University of Liverpool: "Every argument, that, since some change cannot be 'determined' in the sense of 'ascertained', it is therefore not 'determined' in the absolutely different sense of 'caused', is a fallacy of equivocation."[17]

In short, during the half a dozen years that followed the enunciation by Heisenberg of the principle of uncertainty, almost immediately a drastic meaning was grafted on it, a meaning thoroughly philosophical. In that case too the implications of a philosophical position were quickly drawn. Once the scientific inability to measure reality with complete exactitude was given a philosophical garb, it led with philosophical exactness to the inability to grasp and hold reality. Reality's place was taken by chance, not the chance that stands for ignorance, but which stands for a philosophical ghost residing in the shadowy realm between being and non-being. This is why in quantum mechanics no question is ever raised about the sense in which chance is real, that is, part of being, although the quality of being operational is blissfully and unquestioningly attributed to that very same chance.

During the same half a dozen years also the battle lines were firmly drawn. One side was formed by the vast majority of leading physicists, all followers of the Copenhagen interpretation of quantum mechanics, all evasive about questions concerning being, and in that sense all anti-realists. The chief injunction of Niels Bohr, leader of the Copenhagen school, was that all statements about ontology or being must be avoided.[18] The philosophical instructiveness of his writings lies in the consistency with which he obeyed that injunction. The other side was formed by a very small number of prominent physicists—Planck, Max von Laue, Schrödinger, and later de Broglie, with Einstein as their leader. A most distinguished though a very pathetic group indeed. It was nothing short of pathetic that Einstein was lured to a battleground where he could only lose. The ground related to the possibility of devising thought experiments in which the position and momentum could be measured with complete exactness. Einstein assumed a box filled with electromagnetic radiation which was not absorbed by the box because its walls were on the inside perfect mirrors. The box also had a shutter and a clockwork in one of its walls. Once the box was weighed, the clockwork opened the shutter at a preset moment, so that one photon could escape. Thus, at a given time the total mass (energy) of the box diminished, a change that could be ascertained by measuring the box again. Einstein claimed that the experiment made possible the simultaneous measurement of energy and time with complete accuracy, but Bohr was able to show that the weighing process implied a simultaneous measurement of the position and momentum of the scale, a measurement subject to the Heisenberg uncertainty.[19]

No thought experiment immune to that uncertainty can indeed be devised for the simple reason that every measurement ultimately implies the use of light waves. Observation of the results demands the reading of a scale, usually a pointer needle, which can only be done if light is reflected from it into the observer's eyes. But as the light wave is reflected from the needle momentum is transferred to it; the observation of position becomes thereby the simultaneous observation of position and momentum in which Heisenberg's principle sets a limit to the precision that can be achieved. There lies the doubtful source of endless references to the observer's role, references replete with subjectivism, as if the observer as such created reality and nature. Apart from that, even if it were in theory possible to devise a thought experiment with absolute precision, would this in itself be a proof of causality and reality? Obviously not. The very assertion of causality and reality imply a kind of reasoning or rather mental judgment which is very different from statements of mathematical physics.

That such is the case has been amply illustrated by Einstein's inability to speak accurately over many years as he waged his battle against the Copenhagen school. Undoubtedly, Einstein could coin impressive phrases, especially in private. His correspondence with Max Born, which covers over 40 years, is particularly relevant because Born never gave up hope of converting Einstein to quantum mechanics, that is, to its Copenhagen interpretation. Einstein's replies to Born abound in remarkable, at times dramatic lines. In one letter (March 3, 1947), Einstein referred to his little finger as his ultimate proof that quantum mechanics would one day be superseded by a non-probabilistic theory: "I cannot however base this conviction on logical reasons, but can only produce my little finger as a witness, that is, I offer no authority which would be able to command any kind of respect outside my own hands."[20] A pathetic phrase indeed. Not because of the gigantic role attributed to a little finger, but because of the place, the very first place or primacy attributed to logical reasons. Einstein plainly put the emphasis on "logical," not on "reason," let alone on reality.

Similar reflections are in order about Einstein's famous claim that God does not play dice, a claim he made repeatedly in the same correspondence. Take, for instance, the passage from his letter of December 4, 1926, to Born: "Quantum mechanics is certainly imposing. But an inner voice tells me that it is not yet the real thing. The theory says a lot, but does not really bring us any closer to the secret of the 'old one'. I, at any rate, am convinced that *He* is not playing dice."[21] Here too drama goes hand in hand with philosophical poverty. The real thing is not so much reality as perfect calculation, and the God in question is not the One Who Is, but merely someone who can calculate with perfect accuracy and therefore is in no need to play dice. Again there is drama but no philosophical depth in Einstein's remark that, as he put it in a letter to Schrödinger, the Copenhagen people play a dangerous game with reality.[22]

Einstein's public utterances were not much help. They could only convert the believer, that is, those who being realists, needed no conversion. At any rate, conversion is far more than a matter of solid argument. This is why a solid argument must not pretend to imply something which is conversion itself. Einstein made that mistake time and again as he spoke of reality as an object of belief not of knowledge. His famous statement, "belief in an external world independent of the percipient subject is the foundation of all science,"[23] has, for all its longing for reality, a fideistic, or at least a Kantian ring. It is in this fideism toward reality that lies the root of Einstein's inability to make clear over four decades a crucial point in his correspondence with Born, which contains most of his reasonings on the subject, chance versus reality. In battling with Born, Einstein had to take a ground in which no room was left for belief as part of a rigorous argument. But since Einstein made of reality an object of belief, his argument about causality remained hanging in the mid-air of non-reality. So the two, Born and Einstein, talked over 40 years over one another's head until W. Pauli, prompted by Born's frustration, made matters clear and very revealing as well. On learning from Born that Einstein did not wish to continue their correspondence on causality, Pauli wrote to Born that Einstein's principal concern was reality. Causality came second. It made no sense on Born's part, Pauli continued, to dispute causality without facing up to the primary point for Einstein, that is, reality. Such was a perfect clarification of the true order between causality and reality, a clarification which Born did not communicate to Einstein, by then seventy-three and in poor health. Feelings, too, cooled between Einstein and Born. But the most important reason for Born's failure to mention to Einstein Pauli's letter lies undoubtedly in Pauli's revealing comment on the whole dispute. Pauli took the view, which certainly would have outraged Einstein, that questions about reality were as metaphysical and useless as was the concern of medieval philosophers about the number of angels that could be put on a pinhead.[24]

Einstein died without seeing any change in the respective strengths of the two camps. In 1981, a quarter of a century after his death, the situation remains the same. This should be a warning

to those who read too much into the occasional admission of this or
that prominent representative of the Copenhagen school that quantum mechanics may not be the last word in physics. Nothing
changed when Professor Dirac, who first made a name for himself
with a still classic book on quantum mechanics in 1930, declared at
the Jerusalem Einstein Centennial Conference in 1979:

> It seems clear that present quantum mechanics is not in its final
> form. Some further changes will be needed, just about as drastic as
> the changes which one had made in passing from Bohr's orbit to
> quantum mechanics. Some day a new relativistic quantum mechanics
> will have determinism in the way that Einstein wanted. This deter-
> minism will be introduced only at the expense of abandoning some
> other preconceptions which physicists now hold, and which it is not
> sensible to try to get at now. . . . So under these conditions I think it is
> very likely, or at any rate quite possible that in the long run Einstein
> will turn out to be correct even though for the time being physicists
> have to accept the Bohr probability interpretation—especially if they
> have examinations in front of them.[25]

If such a statement, prominent as it is, does not produce a spirited
reaction, and it did not produce any, then it should be easier to
understand why other efforts to break through the walls of the
Copenhagen fortress have been ineffective. Illustrations are the
very weak responses to books and articles written by D. Bohm, J-P.
Vigier and others in the 1950s and 1960s, on behalf of a theory
which permits exact calculations of atomic phenomena, an outcome
which, very wrongly, is taken for proof of causality. Such a theory
is usually referred to as a hidden variable theory, because it implies
that beneath the quantum level deterministic, though hidden factors are at play.[26] Instead of providing a safe foundation for a
philosophy which assumes strict, deterministic causality on the
physical level, the hidden variable theory makes matters even more
confused for those who base their philosophy on the techniques of
mathematical physics. Such seems to be the unintended message of
the famous theorem proposed by John S. Bell in 1964. Any hidden
variable theory, which wants to keep also the statistical predictions
of quantum mechanics, leads, in Bell's words, to the conclusion that
"there must be a mechanism whereby the setting of one measuring
device can influence the reading of another instrument, however
remote. Moreover, the signal involved must propagate simul-

taneously, so that such a theory could not be Lorentz invariant."[27] It may, of course, be true that in Bell's paper not all possible types of hidden variable theories were taken into account. Regardless of this, Bell's theorem is a strong reminder of the impossibility of strict localization by measurement, an impossibility implied in Heisenberg's principle. If this impossibility is then taken in a philosophical sense, and if one still wants to keep the idea of a coherent physical world, then instantaneous communication between instruments is to be assumed. Now if instruments can communicate with one another, then why not minds? Have not minds, under the impact of computer development, become largely taken for natural counterparts of artificial intelligence? Indeed, Bell's paper provided much encouragement to advocates of ESP (extrasensory perception) and TK (telekinesis).[28] The phot*on*, which is the exchange particle of electromagnetic interactions in quantum mechanics, and the gravit*on* of gravitational forces have now a mental counterpart under the name of mind*on*, the presumed exchange particle in ESP and TK, and ultimately in all mental interactions, including any ordinary discourse.

The 'mindon' is one example of the philosophical fruits of the drastic meaning of Heisenberg's principle, a meaning about which one cannot emphasize enough that it is a philosophical meaning going far beyond the purely scientific realm. In addition to the 'mindon', the philosophical history of quantum mechanics provided other telling examples of how the logic operating within a philosophical claim, such as that drastic meaning, unfolds its true nature as time goes on. The last decade witnessed, for instance, the proposition (and a very valid one), that Heisenberg's principle leads to the multiworld theory which states that there are as many worlds as there are observers.[29] If such is the case, the fact that scientists, each of whom has his own individual world, or perhaps even better, is his own individual world, still understand one another, becomes a mystery, or perhaps a sheer miracle. Modern scientists and philosophers would not, of course, propose a miracle for the explanation of that mystery, as did Malebranche and Leibniz, who explicitly considered such a situation. Another example of the same process leading to a philosophically disturbing situation is the principle of "man-centered objectivity," advocated recently by the prominent French physicist Bernard d'Espagnat.[30] No comment is deserved by solipsism, which has for long been recognized as an inevitable implication of the drastic meaning of Heisenberg's principle.

Unfortunately, only on occasion does one find a prominent spokesman of that drastic meaning who is ready to admit that on the basis of that drastic meaning he is not allowed to say that a thief took his wallet, but only that he has the sensation of his wallet having been taken away.[31] What this shows is that the world as articulated in terms of that drastic meaning is a world of philosophical robbery. Those who do not resolutely challenge its proponents lend their support to a situation in which thieves can freely operate without the possibility of ever being apprehended. Such an outcome, in which *to be* and *not to be* are ultimately indistinguishable, is not something to cheer about. Worse, this outcome should have appeared a foregone conclusion half a century ago, at least to those taking a long look at some lines in Bertrand Russell's *Outline of Philosophy,* a book published in 1927, the year when Heisenberg formulated his uncertainty principle. The principle, as I have said, had for some time been in the air, and if anyone could sense the kind of philosophical atmosphere generated by some scientists in 1925 and 1926 it was Bertrand Russell who was writing his book at that time. According to Bertrand Russell: "For aught we know an atom may consist entirely of the radiations which come out of it. It is useless to argue that radiations cannot come out of nothing. . . . Matter is a convenient formula for describing what happens where it isn't."[32] Forty years later the difference between material and strictly non-material was emphatically rejected by H. Margenau: "The quantum mechanical interactions of physical psi fields . . . are wholly non-material, yet they are described by the most important and most basic equations of present day quantum mechanics . . . which regulate the behavior of very abstract fields, certainly in many cases non-material fields, often as tenuous as the square root of probability." No less revealingly Margenau added that the physicist's psi—he had in mind Schrödinger's ψ function—"has a certain abstractness and vagueness of interpretation in common with the parapsychologist's psi."[33]

There are still some who seek a resolution in modern atomic physics to the age-old contrast between freedom and determinism. About speculations, which center on the uncertainty of the motion of electrons in the firing of synapses as a clue to the influence of mind on matter and to the mind's freedom to choose, one remark should suffice. The very same Eddington, who ultimately inspired those speculations, had realized their futility at a very early date.[34] What Eddington failed to recognize was that the ultimate reason behind the failure of such speculations is the difference between quantum mechanics as a science, and the drastic philosophy which its scientific architects, including Eddington himself, erected around that science. The science of quantum mechanics states only the impossibility of perfect accuracy in measurements. The philosophy of quantum mechanics states ultimately the impossibility of distinguishing between material and non-material, and even between being and non-being. Physicists who fail to realize what this means for their science should remind themselves of a remark of James R. Newman, for many years the editor of *Scientific American* and always full of admiration for the work of physicists: "The more creative physicists have in recent years cultivated philosophy. They are usually disinclined to admit to this weakness. But there is no escape, even if it be only to embrace anti-philosophical philosophies. For the physicist has come to realize that if he throws philosophy into the fire, his own subject goes with it."[35]

At any rate, if it is impossible to distinguish between being and non-being, then efforts to say anything about freedom and determinism become utterly meaningless. Of course, scientists incuding the leading spokesmen of the Copenhagen school, would never admit that they were not truly free as they searched for and made their great discoveries. For if they were not free, what is the ground for their receiving awards and Nobel Prizes? But if they admit that they were free, their philosophy of quantum mechanics must face up to at least one certainty on the most fundamental level of existence. That they do not see this contradiction provides only one more example of the truth of Einstein's statement, "The man of science is a poor philosopher."[36] He was indeed one, in the sense that he could never articulate his good philosophical instinct. He failed to write even a short article on his recognition that science was merely a refinement of common sense. He did not even suspect that Scottish philosophers, who in the second half of the eighteenth century made the term "common sense" fashionable,[37] did not steer philosophical development toward that realism which he wanted to vindicate. But Einstein at least recognized that the science of physics entitled no one to sit in judgment over the question of freedom versus determinism. Although once more his statements were not articulate, they were certainly dramatic. It is not the uranium but the heart of man that should be purified, he said to a journalist in the wake of Hiroshima.[38] About the same time he admitted to one of his biographers that he never derived a single ethical value from physics.[39] His most revealing statement in this connection is from his correspondence with Born. Still during World War II Born urged Einstein to propose the formation of an International League of physicists to prevent the turning of physics into a tool of global destruction. Einstein replied: "The medical men have achieved amazingly little with a code of ethics, and even less of an ethical influence is to be expected from pure scientists with their mechanised and specialised way of thinking."[40]

Einstein's description of scientific thinking as mechanized may seem to contradict his own repeated statements that there was no logical, that is, mechanical way to discoveries. Discoveries, great conceptual novelties, are few and far between in science. For most of the time most scientists reason within an already given conceptual framework. In view of the univocal character of their whole subject matter, or the quantitative aspects of entities and their interactions, their reasonings often resemble the operation of a logic machine. Yet even then reasoning as a free act of understanding is very different from a purely mechanical operation. Interestingly, it was at a time, during the second half of the nineteenth century, when classical deterministic physics seemed to come within the explanation of all physical interaction, that leading physicists (Maxwell, Lord Kelvin, Helmholtz, and many others) stressed the incompetence of physics in matters relating to freedom.[41] A major echo of that emphasis came in our century from none other than A. H. Compton whose work was indispensable for the formulation of Heisenberg's principle. In a lecture series given at Yale in 1934 Compton declared that one's inner conviction to move one's finger at will carried greater and more immediate evidence than all the evidence of the laws of physics, and if freedom and physics were ever to be found in conflict physics was to be corrected and not our freedom to be doubted ever so slightly.[42]

Proponents of the drastic meaning of Heisenberg's principle still have to face up to some problems with all the seriousness required by the issues at stake. One problem is the definition of chance. Do they mean something ontological or something which is merely a mathematical device? If they mean the latter, they should ask themselves whether there is a mathematical theory of randomness which would not include at least one, subtly concealed non-random parameter in the ensemble. If they have something ontological in mind, they should ask themselves whether, within their perspective, chance can be anything but a negation of ontological causality. In that case they should ponder the problem of non-being as the cause of something, that is, a being. For such is ultimately the problem of a physical interaction in which either the effect comes into being without a cause, or it contains a surplus with respect to its cause. In a more specific sense they should face up to the problem of absolute chance and absolute chaos. It may help them if they recall that this problem was not first posed by modern quantum mechanics and Heisenberg's principle. The problem was much discussed in connection with natural selection as postulated by Darwinian evolution. Unlike T. H. Huxley, many Darwinists were inclined to take that selection for a purely random process. Had not such been the case, the famed logician, Charles S. Peirce, would not have, almost exactly a hundred years ago, begun his comments on Darwinism with the following phrase: "A truly evolutionary philosophy of nature would suppose that in the beginning—infinitely remote—there was a chaos of unpersonalized feeling, which being without connection or regularity would properly be without existence."[43]

That the passage from non-existence to existence on the basis of perfect chaos or pure chance bothers few scientists and philosophers today tells much of the true measure of their sensitivity to what really matters. The fearless logic, with which the implications of the drastic meaning of Heisenberg's principle have been drawn, has not been matched by a fear of that logic which is the art of going wrong with confidence. Admiration is certainly owed to a thinker, say, a St. Augustine, who after much pondering on such a deep problem and immediate experience as time declares: When you don't ask me about it I know what it is; when you ask me, I don't know. Chance may be a problem, though hardly deeper than any branch of probability calculus. Chance is certainly not an experience, not even in atomic physics. Proponents of the drastic meaning of Heisenberg's principle failed to come up with the kind of chance which Schrödinger once pointedly described as "intelligible chance."[44] About chance in that drastic or counter-ontological meaning it hardly can be claimed that one knows what it is when not asked about it. The fact, an often observed fact, however, is a baffled silence given for answer when the debate is shifted from technicalities covering up that drastic meaning to the plain and blunt question: What *is* chance?

If that drastic meaning of chance is vindicated on the ground that knowledge is to be suspended about everything which is not directly observable, only the realm of philosophical poverty and insensitivity is extended beyond limit. Instead of rehashing the old though ever timely subject of universals, let me rely on the principle that one illustration is worth a thousand definitions. Claude Bernard was a famed student of life processes all his life both as an experimentalist and as a philosopher. He studied at length the question whether life was to be explained in a mechanistic or in a vitalistic framework. To someone pressing him on this point, he once gave the terse reply: I have never observed life. He, of course, would have never stated that he did not know life, in spite of the fact that he never observed it. To know life, and to know an immense range of entities, stretching from mere matter through organic life to men (including the physicists of the Copenhagen persuasion), much more is needed than mere observation. Much the same surplus is contained in our knowledge of physical interactions versus measurements in physics. Knowledge of that interaction existed long before physics and physicists. That knowledge relied of course on a rough, commonsense estimate of the equality of effect with cause. To make the certainty of that knowledge an exclusive function of measurements with no uncertainty involves not so much a scientific impossibility as an elementary error in logic. Ancient Greek philosophers gave it the name, μετάβασις εἰς ἄλλο γένος, the favorite technique of those who thrive on confusion and also of those who somewhat innocently dupe themselves. In this age of quantum mechanics and of lucky and unlucky shots of all sorts, one may prefer the name "philosophical quantum jump," a most fatal jump indeed if judged by its consequences for our understanding of measurements in physics, interactions in nature, chance, determinism, to say nothing of such far deeper and more crucial topics as freedom and reality itself.

---

[1]TIME, April 27, 1981, p. 79.
[2]*Lettres philosophiques* (1734), in *Oeuvres complètes de Voltaire*, ed. L. Moland (Paris, Garnier Frères, 1877-85), vol. XXII, p. 130.

[3]Laplace first made this statement in his *Théorie analytique des probabilités* (1812). The statement gained wide currency through its insertion into Laplace's popular exposition of the same topic, *Essai philosophique sur les probabilités* (1814). See its English translation from the sixth French edition, *A Philosophical Essay on Probabilities*, by F. W. Truscott and F. L. Emory, with an introductory note by E. T. Bell (New York, Dover, 1951), p. 4.

[4]Article "Atomes" in *Dictionnaire philosophique* (1764). See *Oeuvres complètes de Voltaire*, vol. XVII, p. 478.

[5]*Wallenstein's Death*, Act II, Scene 1, lines 943-44. Schiller may have just as well relied on the dictum of Lessing, a leader of the German Enlightenment: "Nothing under the sun is ever accidental" (*Emilia Galotti*, IV). Alexander Pope, world-renowned British spokesman of rationalist optimism, attacked chance with no less resolve: "All chance, direction which thou canst not see" (*An Essay on Man*, I).

[6]For the printed text of those reminiscences, see *The Life and Letters of Charles Darwin*, edited by F. Darwin (New York, Basic Books, 1959), vol. 1, pp. 553-55.

[7]First published in 1970. Monod explicitly refers to Heisenberg's uncertainty principle as he sets forth the notion of "pure chance, absolutely free but blind" and contrasts it with Laplace's statement. See *Chance and Necessity: An Essay on the Natural Philosophy of Modern Biology*, translated from the French by A. Wainhouse (New York, Vintage Books, 1971), pp. 112-15.

[8]"Über den anschaulichen Inhalt der quantentheoretischen Kinematik und Mechanik," *Zeitschrift für Physik* 43 (1927), p. 197. Heisenberg's paper was received by the editor on March 23 and appeared in early June.

[9]He did so in his Gifford Lectures, delivered at the University of Edinburgh, January-March 1927. See *The Nature of the Physical World* (Cambridge, University Press, 1928), p. 294.

[10]See *Nobel Lectures. Including Presentation Speeches and Laureates' Biographies. Physics* 1922-1941 (Amsterdam, Elsevier, 1965), p. 301.

[11]*Mathematical Foundations of Quantum Mechanics*, translated from the German by R. T. Beyer (Princeton, Princeton University Press, 1955), p. 327.

[12]The debate of Dirac and Heisenberg took place at the Solvay Congress, Bruxelles, October 24-29, 1927. The papers and procès-verbaux of the Congress are contained in *Electrons et photons* (Paris, Gauthier-Villars, 1928). See especially pp. 261-63.

[13]"The Decline of Determinism," *The Mathematical Gazette* 16 (1932), p. 74.

[14]"The New Vision of Science," *Harper's Magazine* 158 (1929), p. 450.

[15]*The Nature of Physical Reality* (New York, McGraw Hill, 1950), p. 418.

[16]*Nature,* March 26, 1927, p. 467.

[17]*Nature,* Dec. 27, 1930, p. 995. Turner attached his remarks to the assertion made by the Nobel laureate physicist, G. P. Thomson, in his book, *The Atom* (London, T. Butterworth, 1930, p. 190), that "physics is moving away from the rigid determinism of the older materialism into something vaguely approaching a conception of free will." The assertion is a perfect example of the kind of vagueness incompatible with philosophy.

[18]See for details my Gifford Lectures, *The Road of Science and the Ways to God* (Chicago, University of Chicago Press, 1978), pp. 211-12.

[19]For description, discussion, and diagrams, see N. Bohr, *Atomic Physics and Human Knowledge* (New York, John Wiley, 1958), pp. 32-66.

[20]*The Born-Einstein Letters: Correspondence between Albert Einstein and Max and Hedwig Born* (New York, Walker and Company, 1971), p. 158.

[21]Ibid., p. 91. For other uses by Einstein of the same phrase, see ibid., pp. 149 and 199.

[22]Letter of December 22, 1950 of Einstein to Schrödinger in *Letters on Wave Mechanics: Schrödinger, Planck, Einstein, Lorentz,* edited by K. Przibram, translated with an introduction by M. J. Klein (New York, Philosophical Library, 1967), p. 36.

[23]"Maxwell's Influence on the Development of the Conception of Physical Reality," in *James Clerk Maxwell: A Commemoration Volume 1831-1931.* Essays by Sir J. J. Thomson et al. (Cambridge, University Press, 1931), p. 66. The continuation of the statement, "But since our sense-perceptions inform us only indirectly of this external world, or Physical Reality, it is only by speculation that it can become comprehensible to us," reveals Einstein's inability that, for all his longing for realism, he could not free himself of the shackles of Kant's philosophy which he imbibed as a teenager.

[24]*The Born-Einstein Letters,* pp. 221 and 223.

[25]The statement saw print only through its having been quoted by R. Resnick, "Misconceptions about Einstein: His Work and his Views," *Journal of Chemical Education* 52 (1980), p. 860.

[26]For a still very useful and amply documented discussion, see *A Survey of Hidden-Variables Theories* by F. J. Belinfante (Oxford, Pergamon Press, 1973).

[27]"On the Einstein-Podolsky-Rosen Paradox," *Physics* 1 (1964), p. 199.

[28]As can readily be gathered from *The Roots of Coincidence,* by A. Koestler (London, Hutchinson, 1972), a book devoted to the support of the claim that modern physics validates experiments on ESP and TK. It tells much of Koestler's philosophy that he ends with an advocacy of a world-mind into which all individual minds are diffused to the extent of losing their identity. The saving of free will on the basis of modern physics exacts indeed a very heavy price.

[29]For a brief account, see the concluding section, "Many-World Theories,"

in M. Jammer, *The Philosophy of Quantum Mechanics: The Interpretations of Quantum Mechanics in Historical Perspective* (New York, John Wiley, 1974), pp. 507-21. For a criticism of Jammer's philosophically facile presentation of those theories, see my *The Road of Science and the Ways to God*, p. 411.

[30]In a letter to the editor of *Scientific American* 242 (May 1980), pp. 8-9. No more philosophical merit is contained in the reification of quantum states through which V. Weiskopf believed to have vindicated external reality in his criticism (also in a letter to the same editor, ibid., pp. 6-7) of d'Espagnat's article, "Quantum Theory and Reality," *Scientific American* 241 (Nov. 1979), pp. 158-81.

[31]Based on a private dispute of this author with a Nobel laureate physicist.

[32]Quoted from the American edition which has the title *Philosophy* (New York, Norton and Norton, 1927), pp. 156-59.

[33]"ESP in the Framework of Modern Science," in *Science and the ESP*, edited by J. R. Smythies (London, Routledge and K. Paul, 1967), p. 209.

[34]Eddington, who suggested in 1934 in his lectures at Cornell University (*New Pathways of Science*, Cambridge, University Press, 1935, p. 88) that calculations of the width of uncertainty may be an indication of the "measure" of human freedom, declared such suggestions to be nonsensical in his *The Philosophy of Physical Science* (London, Macmillan, 1939), p. 182.

[35]Such is the conclusion of Newman's long review of D. Bohm's *Causality and Chance in Modern Physics* (1957) in *Scientific American* 198 (Jan. 1958), p. 116.

[36]"Physics and Reality" (1936), in *Out of My Later Years* (New York, Philosophical Library, 1950), p. 59.

[37]The emphasis laid by T. Reid and his followers on the instinctiveness of common sense lent ultimately support to theories of knowledge steeped in emotionalism if not plain irrationalism. See for details, E. Gilson, *Réalisme Thomiste et Critique de la Connaissance* (Paris, Vrin, 1939), pp. 14-22.

[38]In an interview with M. Amrine, *The New York Times Magazine*, June 23, 1946, pp. 42-44.

[39]P. Michelmore, *Einstein: Profile of the Man* (New York, Dodd, 1962), p. 251.

[40]*The Born-Einstein Letters*, p. 148.

[41]See Chapter ix, "Physics and Ethics," in my *The Relevance of Physics* (Chicago, University of Chicago Press, 1966).

[42]*The Freedom of Man* (New Haven, Conn., Yale University Press, 1935), p. 26.

[43]*Collected Papers of Charles Sanders Peirce*, edited by Charles Hartsthorne and Paul Weiss (Cambridge, Mass., Harvard University Press, 1931-35) vol. VI, 33. The rest of the passage is no less expressive of the true logic of absolute or pure chance as implied in typical evolutionary

theory: "This feeling, sporting here and there in pure arbitrariness, would have started the germ of a generalizing tendency. Its other sportings would be evanescent, but this would have a growing virtue. Thus, the tendency to habit would be started; and from this, with the other principles of evolution, all the regularities of the universe would be evolved. At any time, however, an element of pure chance survives and will remain until the world becomes an absolutely perfect, rational, and symmetrical system, in which mind is at last crystallized in the infinitely distant future." Compared with this dash in a few lines from the non-being of pure arbitrariness to an absolutely perfect system while pure chance remains always at work, Aristotle's often decried one-page-long derivation on a priori grounds in his *On the Heavens* of the shape and structure of the universe should appear a very sober enterprise.

[44]E. Schrödinger, *What is Life and Other Scientific Essays* (Garden City, N.Y., Doubleday, 1956), p. 199. Schrödinger was right in seeing "intelligible chance" at work both in Boltzmann's statistical theory and in genetic mutations because both imply several clearly defined parameters.

Note added in proofs. The result, as reported in *Physical Review Letters* (Aug. 17, 1981, pp. 460-63), of the experiments of A. Aspect et al., on the linear polarization correlation of photons emitted in a radiation cascade of calcium, strongly suggests the incompatibility of quantum mechanics and of all hidden variables theories which retain the relativity principle that no signal can travel faster than light. The result can, however, have no bearing on the question of causality if it is true, as was argued throughout this paper, that causality is not a function of the possibility of measuring with absolute precision. That the result is being played up in a counter-ontological sense is as much part of the contemporary philosophical malaise as is the taking of hidden variables theories as equivalent to ontological causality because they imply *measurably* exact localization in space. Only somersaults in logic can make "measurably exact" appear identical to "real."

2

From Subjective Scientists
to Objective Science

In surveying a topic, it is customary and very useful to turn first
to the index cards in the subject catalogue of a good library. Unlike
museums, which catalogue their *objets d'art,* good libraries are now
busy developing subject catalogues, although libraries are built for
housing such plain objects as books. This obvious ambivalence in
semantics should suggest something of the dilemma inherent in the
relation between subject and object, subjective and objective. The
dimensions of this dilemma are readily shown by the index cards
grouped under the headings, object and objectivity. Much the same
would, of course, be the result if one consulted the index cards on
subject and subjectivity. This symposium is, however, on objectivity and therefore let me recall by way of introduction what my
perusal of some index cards in the subject catalogue[1] taught me in a
short hour about object and objectivity.

Paper presented at the 3d International Humanistic Symposium in Athens
and Pelion, 1975; reprinted with permission from its *Proceedings,* Athens,
1977, pp. 314-30.

Among the cards there were quite a few on objectivity in education and psychology, a fact which should not be surprising. When
claims are exaggerated, perplexities abound and publications pro-
liferate. The perplexities can reach as far as to trouble the American Institute of Ceramic Engineers, not a widely known group to
be sure. Engineers are supposed to be matter-of-fact people, and
what could be more objective than doing ceramics? But as the cards
told me, those engineers had to organize—and with the help of the
National Science Foundation—a study on objective criteria in
ceramic engineering education. That was in the 1960s, the good old
times when the same Foundation could spend its many millions of
dollars in a manner that has recently prompted if not an objective,
at least a Congressional, investigation. So much about education
and objectivity.

As to psychology, one index card referred to a recent, and fairly
massive, book the title of which, "objective personality assess-
ment," was immediately put in its proper subjective light by the
subtitle, "changing perspectives." Apart from limitations on our
time this might have been a good reason for not discussing education and psychology in this symposium on objectivity. Those interested in our first day's topic, classical philology, might wish to hear
about a book on objective methods for testing the authenticity of
literary works. The book was written about some comedies of
Lopez de Vega. The times he lived in and wrote about were the heyday of classical studies which on occasion amounted to mere comedies. Galileo might have meant precisely this when he remarked
that in the humanities, which at that time were mainly the study of
classics, there was neither truth nor falsehood.[2] What he meant,
rightly or wrongly, was that such studies were purely subjective.
As to philosophy, the topic of our second day, a title among the
index cards referred to objectivity according to Kant, an obviously
subjective study, since by Kant's own definition nothing could be
known about the object as such, or the *Ding an sich*. Concerning
sociology, or the topic of our fourth day, one cannot help thinking
today of Marxism, and the cards once more lived up to my already
heightened expectations. One card referred to a recent book on
"dialectics and the objectivity of praxis," but one wonders whether
praxis can be anything but subjective when trapped in the laby-
rinths of dialectics. Although we have dropped plans to discuss
theology, mention might be made of a card which referred to God's

objectivity being on trial, a story, if I may add, as old as the serpent's success in lulling Eve into thinking that God was not as objective as He claimed Himself to be. It was somewhat surprising to see a title on objectivity in accounting, but it is perhaps a reflection on our times that objectivity may no longer have close ties with plain honesty.

Finally, to complete my account of index cards, I must mention one which informed me about a symposium held in Belgium on objectivity in the exact sciences.[3] It deserves to be mentioned not only because of the timeliness of our undertaking, but also because of the last line of the proceedings of that symposium. There high praise was given to the idea of the *mathesis universalis* as the supreme embodiment of objectivity.[4] Curiously, the batch of index cards I went through contained no reference to objectivity in mathematics, which brings me to the topic of this third day of our symposium. Happily for mathematics, only mathematicians write about it, and they had already reached the conclusion two generations ago that their subject—I mean object—demands no objectivity, that is, reference to the objective, external world, but only consistency.[5] Hence the absence of cards on objectivity in mathematics. This lack of objectivity in mathematics does not, however, justify Bertrand Russell's remark that "mathematics is the subject in which we never know what we are talking about, nor whether what we are saying is true."[6] Joshua Willard Gibbs, the famous American mathematical physicist, was much more objective when he remarked about the same time, the turn of the century, that a "mathematician may say anything he pleases, but a physicist must be partially sane."[7] What Gibbs meant by "being sane" was being objective. By not being objective one can still perhaps be sane, but only, it seems, if one is a mathematician.

From mathematics it is but one and a natural step to the exact sciences, especially to physics, which had earned for some time the reputation of being the paragon of objectivity. The subject catalogue on objectivity was once more a mirror of the actual climate of thought, but a mirror which resembled mirrors in amusement parks: you look into them and you see yourself standing on your head. Yet, it was not a librarian's trick, or mistake, to place among cards on objectivity a card of a book published in 1967 by Professor Scheffler of Harvard, a book whose title is *Science and Subjectivity.*[8] The publication of that book was a symptom, a sudden awakening to the true magnitude of a movement which claimed that science, long the paragon of objectivity, could perhaps be exact but not at all objective.

The beginnings of that movement were rather innocuous. The unlocking of atomic energy catapulted scientists into a political preeminence, which immediately made them open to public criticism. As long as scientists remained inside their laboratories, their image of being objective in all circumstances was above suspicion. Once, however, they had accepted the role of political advisers, they turned out to be no less subjective than politicians are. This is what prompted James F. Byrnes, the U.S. Secretary of State, to remark: "In this age it appears every man must have his own physicist."[9] Soon afterwards there followed the agonizing disputes about the wisdom of making atomic bombs and once more scientists were cut down to a very common, objective size, precisely through the proofs of their subjectivity. These were the 1950s which saw the publication of such books as *Science Is a Sacred Cow,*[10] to mention only one title that could have hardly found a favorable echo a generation or so earlier. Since the present always invites the past, before long it became a favorite pastime to speak of Copernicus, Kepler, and Galileo as sleepwalkers and to understand Newton in terms of Freud.[11] But even when the history of science was not cultivated with some patently subjective aim, or objective, in mind, the question of the subjective could no longer be avoided. In that tremendous upsurge of interest in the history of science that started in the early 1950s and is still growing it became a tenet—and a fairly objective one—that in order to understand what scientists objectively think, do, and say, one must probe into the subjective roots of their thoughts, deeds, and statements.[12]

Scientists can be very subjective, and awareness of this is never a luxury. In this age of ours, when the tools created by science embody awesome potentialities, it can never be remembered strongly enough that a scientist is as prone to foibles and prejudices as any other human being.[13] Like other humans they, too, have their subjective limitations and peculiarities. Yet unlike others, scientists seem to communicate with one another at least in matters scientific with a much greater effectiveness than other professional groups. The usual explanation of this is that either their subject matter or their method raises scientists above their subjectivity. This explanation is not to be taken lightly, but it is not so much an explanation as a statement. It is simply the recognition of the relative ease of communication among scientists, a fact which is undeniable. From this it is tempting to make the inference that there exists something supra-subjective, or simply objective, in or behind scientific practice, but such an inference is more of a vague intuition than a carefully reasoned discourse. At any rate, such an inference leaves open the question whether it is the object of scientific work, or the scientific method dealing with that presumed object, which secures that supra-subjective communication among scientists.

It would now be natural to meet head-on the question of object and objectivity. Such would primarily be a task for a philosopher, but only for a philosopher who had not yet parted with his common sense; for object, objective, and objectivity can be had only on such a ground, a point to be discussed later. What can already be stated here is that common sense requires careful attention to the common scene, and what is, one may ask, more common in the modern intellectual arena than are science and scientists? It will certainly show good sense on the part of the philosopher if he notes that the most creative minds in exact science have always insisted, and emphatically so, that science cannot exist without a scientist's firm belief in the existence of an objective world. Among these most creative minds were certainly all the great figures of classical physics, Copernicus, Kepler, Galileo, Newton, Euler, Lagrange, Faraday, Helmholtz, and Maxwell.[14] The twentieth-century scene is more complex. That belief in an external, objective world was the very foundation of the scientific spirit was a tenet vigorously upheld by Planck and Einstein, by far the two most creative figures in modern physics. It should seem very significant that both came to espouse this tenet by first rejecting, after a long reappraisal, the philosohy of Ernst Mach, for whom subjective sensations were the only objective data.[15] Particularly revealing in this respect is the fact that both Planck and Einstein lived their younger years as ardent followers of Mach and it was their creativity in science which persuaded them that no solid meaning could be given to the scientific enterprise if belief in an external, objective world was not upheld vigorously and unconditionally.

Beside Planck and Einstein, the giants, there are many lesser giants in modern physics, such as Bohr, Born, Heisenberg, Schrödinger, Dirac, de Broglie, and Wigner, to name only a few.

With them the assertion of an objectively existing world—the obvious ground for objectivity—shows a broad spectrum. It is much wider on the side bordering on agnosticism than on the side relating to realism.[16] All these men started their work as physicists when Ehrenfest remarked that teaching mathematical physics in its present confusion had given him a glimpse into Hegelian dialectics, a "succession of leaps from one lie to another by way of intermediate falsehoods."[17] With the formalization of quantum mechanics by Heisenberg and Born, the lies seemed to have been eliminated, but only at the price of reducing the objective to the strictly observable. This position was so quickly and so widely accepted as to create a school, the Copenhagen interpretation of quantum mechanics. Obviously, the position in question implied far more than mere theoretical physics. With an eye on the Copenhagen school, Einstein noted as early as 1935 that most contemporary physicists were speaking as if they were followers of Mach.[18] Indeed, the so-called empirically observable was in the Copenhagen interpretation equivalent to sensations as interpreted by Mach and with him no room was left for an objectivity rooted in an objective world.

Such a development was certainly a major paradox as far as the philosophy and history of science were concerned. From the earliest times the principal aim of scientists consisted in formulating a method which would secure objectivity, that is, the ability to reach the objectively existing world. Quantum mechanics, so it seemed, demanded only consistency, not objectivity, and as a result, physicists could appear to become like mathematicians, who could say whatever they wanted to say. Some physicists did indeed wade into philosohy as if they had been less than partially sane.[19] Philosophically, quantum mechanics began to give at times the impression of resembling the insanity of too much success. For in quantum mechanics it was the very success of the method which seemed to justify the denial of the validity of statements about an objective world. It was in such a milieu that were born those recent interpretations of science which in turn prompted Scheffler and others to sound the alarm.[20] The milieu was scientific on the surface but heavily philosophical underneath. Thomas Kuhn, for instance, lived in that milieu before he created so much stir with his paradigms, partly because he claimed that paradigms, or mental constructs, were at times constitutive of nature itself.[21] This could only mean, philosophically that is, that the subjective produced the objective, if the latter existed at all. It was as a student of physics at Harvard in the 1940s that Kuhn could not help becoming exposed to current philosophical interpretations of exact science.[22] What struck him was the great difference of these interpretations from the spirit in which science was cultivated by Galileo, Newton, and their followers. Clearly, none of them subscribed to a philosophy of sensationism, or of empiricism, much in vogue nowadays, a vogue which prompted Bertrand Russell to remark that empiricism was a fashion. What is, however, a fashion, if not a universal foible, or speaking more philosophically, an objectively shared subjectivism? To quote Russell: "Many issues are decided by many people on a basis of party spirit, not [on the basis] of detailed examination of the problem involved. In particular, whatever presents itself as empiricism is sure of widespread acceptance, not on its merits, but because empiricism is the fashion."[23]

If there is an objective truth about fashions, it may perhaps be the fact that any fashion is a disease which is never cured but only replaced. Actually, when a fashion is purported to be cured by another fashion, the remedy is often worse than the disease itself. Yet this vicious process is not without benefits. When a disease becomes critical, illusions can no longer be entertained. Hence the sudden rush to the alarm bell, although in this case it has been sounded largely by those who on the basis of their professed philosophy have no logical ground for stepping into the breach on behalf of objectivity and objective science.[24] The reason for this can only be seen if it is recalled that the vanishing of an objective world within the commonly accepted philosophical interpretation of quantum mechanics has far less to do with quantum mechanics than with philosophy in general and with the philosophy of scientific method in particular. The Copenhagen interpretation of quantum mechanics is the major specific source of modern doubts about objectivity that have to do with science, because the rejection of objectivity by existentialism is predicated on its wholesale rejection of science itself.[25] The Copenhagen interpretation is, however, not so much a derivative of quantum mechanics as a fashionable dress forced on it. It is a dress fashioned by scientists who did not care, in a rather unobjective fashion, to take a long look at the philosophical roots of their philosophy of scientific method, a philosophy which left logical room only for method but not for objectivity and objective world.

This long look has to be such not only because of the seriousness of the matter but also because the history of concern for objectivity is very long. A long history, always a delight for the historian, can be a tedium for the philosopher, but only if he forgets that the first phase of a long story usually contains the logic of subsequent developments. The first phase of our story takes us back to Hesiod, the first to try to give an objective explanation of something which appeared to him, and not without good reason, the most objective phenomenon. The phenomenon in question was strife. Little is known about the political history of the Greeks of Hesiod's time, but it must have been full of strife if the centuries[26] immediately following can serve as a clue to Greek prehistory. The same, of course, would be true of any other nation, tribe, or group. Strife is a universal human phenomenon and nothing is more natural than to project social and personal strife into nature, whose various parts, elements, and forces seem to be in constant strife. One can therefore understand Hesiod, who put a personal factor, a god, behind each phenomenon of nature. This is why the cosmogony of Hesiod is a theogony,[27] an explanation of the objective in terms of the subjective, an obviously extremist position. It was so extreme as to produce an almost equally extreme reaction on the part of the Ionians. That there was strife in nature and nothing but strife, this the Ionians did not doubt, and much less was this doubted by Empedocles and Heraclitus. But that there were as many gods as there were strifes in nature was a different matter. Empedocles and Heraclitus reduced the number of gods to two, love and hatred. The Ionians left no god at all. They eliminated the subjective to save the objective. Scientifically speaking it was a tremendous progress to speak of earth, or water, or air, or fire, instead of gods. But there was a fly in the ointment. No agreement could be secured as to what was really objective, the earth, the water, the air, the fire, or all of them, or perhaps none of them. However, the successive predominance of these elements could easily be established, and on the basis of that sequence phenomena could be predicted. But sequence was one thing, object or objective another, unless the latter was reduced to mere sequence.

What the Greeks, and in particular Socrates, perceived was that if one followed the Ionians, and Anaxagoras in particular, one could not account for the most fundamentally real of all human experiences, the experience of acting freely and for a purpose. Freedom and purpose also meant sequence but not an inevitable one, yet a very real sequence. To save that real sequence Socrates proposed a new physics—the story is well known from the *Phaedo*—a physics in which everything was based on purpose, or rather volition.[28] The essence of that physics was that method and object, if not reality, were thematically identified. The method was the subjective introspection which informed about purpose, the object was the purposeful striving in nature. As to reality, it was not exclusively identified with purpose, but a cloud was unwittingly brought over inert reality by this overemphasis on purpose. All this meant a return to the position of Hesiod but on a level more abstract as can be seen in Aristotle's *On the Heavens* and *Physics*, the systematizations of Socrates' ideas on what true physics ought to be.

The polemical parts of Aristotle's foregoing works were concerned with the reaction to Hesiod, a reaction which the atomists carried far beyond the position of the Ionians. The atomists unfolded the essence of the position of the Ionians by postulating the existence of atoms which were the embodiments of those qualities about which an impersonal sequence could readily be predicated. These qualities were later called the primary qualities. Like Socrates, the atomists, too, firmly believed in objective reality. However impervious to the senses, the atoms were believed to be real, nay reality itself. But like Socrates, the atomists paid little attention to the relation between objective and real. Like Socrates, they were preoccupied with the connection between the objective and the method. While Socrates reduced to purpose both method and the objective, the atomists defined both method and the objective in terms of primary qualities wholly devoid of any purpose or volition. As in the Socratic-Aristotelian case, here, too, the procedure was a boomerang that was to strike back at reality itself. As he reviewed the position of the atomists, the biologist Galen made in the second century the prophetic remark that the very same arguments by which doubt was cast on the reality of secondary qualities were to undermine conviction about the reality of primary qualities as well.[29]

The remark indeed anticipated the gist of modern philosophy from Descartes and Galileo to Mach and Carnap. In speaking of Descartes and Galileo one should never forget that as philosophers they wanted to be scientists and, as scientists, they never stopped philosophizing and in a most systematic manner at that. Again, when we turn to Locke and Hume, it is crucial to recall that they aimed at formulating a philosophy as rigorous and objective as the science of Newton. The same is true of Kant, Mill, and Mach. The immediate chief result of all these philosophies was the vanishing of conviction about the real. Hume was particularly explicit in stating that by looking at stars one did not get in touch with stars as extra-personal realities.[30] The only reality was the observer's sensation, a position to which Herschel had already taken an explicit and indignant exception.[31] A Humean might scoff at that but even he must admit that it is rather unlikely to expect Herschel to look through his telescope with one eye, read with his other eye the *Treatise* of Hume, and hear from him that he was seeing no stars but had only sensations. Almost two centuries later Einstein remarked that Mach, the sensationist, should have simply rejected the notion of external reality.[32] This remark of Einstein aimed not so much at Mach the philosopher, as Mach the scientist. For Einstein was above all a scientist, a giant among scientists, and it was his scientific genius that taught him the fact that the notion of the real was a more fundamental notion than the notion of the scientifically objective.[33] Moreover, Einstein also perceived that it was impossible to secure reality to what is known as the scientifically objective, if the latter was made equivalent to reality. Einstein also realized that access to the real could not be made through the channels of positivism. Logical positivism, the ultimate refinement of the identification of objective and real with primary qualities, could in the last analysis offer nothing but words. What Russell noted in his inimitable crisp style about Neurath's position is true of the whole of logical positivism and even of the whole of the Cartesian, Galilean epistemology and of its dead-end fully unfolded by Hume and Mach:

> Neurath's doctrine, if taken seriously, deprives empirical propositions of all meaning. When I say 'the sun is shining' I do not mean that this is one of a number of sentences among which there is no contradiction; I mean something which is not verbal, and for the sake of which such words as 'sun' and 'shining' were invented. The purpose of words, though philosophers seem to forget this simple fact, is to deal with matters other than words. If I go into a restaurant and order my dinner, I do not want my words to fit into a system with

other words, but to bring about the presence of food. I could have managed without words, by taking what I want, but this would have been less convenient. The verbalist theories of some modern philosophers forget the homely practical purposes of every-day words, and lose themselves in a neo-neo-Platonic mysticism. I seem to hear them saying 'in the beginning was the Word', not 'in the beginning was what the word means'. It is remarkable that this reversion to ancient metaphysics should have occurred in the attempt to be ultra-empirical.[34]

Instead of being astonished with Russell that this development was "remarkable," one should simply recognize that this development was utterly logical. Once one attempts to reduce to the so-called objective all that is real, both reality and objectivity are replaced by a neo-neo-mysticism, which is far worse than whatever mysticism Plato might have had. This new mysticism consists only of words that have no relation to reality and therefore they cannot be objective. Russell's remark about that "old metaphysics" is even more misleading than his excoriation of Plato. In fact, Russell's facile brushing aside of "old metaphysics" may very well deprive one of the sole method of recovering reality for philosophical thinking. I must add this because even Russell is known to have remarked that he was not a philosopher to the extent of living by its precepts the every-day, commonsense life. Rational man, as Russell defined him, could not take for certain the rising of the sun next morning, but in that sense Russell preferred not to be a rational man, or a philosopher in short.[35] As a philosopher Russell recognized that empiricism does not lead to reality. As to Kant, the chief protagonist of critical idealism, Russell aptly remarked that Kant was a mere "misfortune" in the history of Western thought.[36]

Idealism and empiricism are extremes and they should appear to be such both historically and analytically. Between extremes one instinctively looks for a middle ground and this middle ground can easily be spotted, though not, of course, if with Russell one eliminates all "old metaphysics." It was within that vast realm of "old metaphysics" that it was perceived long ago that knowledge puts one in immediate contact with reality. Like any other statement, this proposition, too, can be formulated with or without a large apparatus of conceptual refinement. But even with the whole apparatus given, the heart of the matter boils down to the almost existential assertion that man through his consciousness is always in touch with a reality existing independently of him.[37] This consciousness of his does not create reality, but since it is never without some sensory content, he must believe that this consciousness leads him to a reality existing independently of him, or else all his sensations will ultimately prove illusory. The whole history of modern (and ancient) philosophy shows that when any other starting point is taken in epistemology the truth of reality, including that of external reality, cannot be secured.[38]

Without securing reality, the so-called scientific objectivity or empirical objectivity has no secure foundation. The whole development of modern science brings out the same lesson. The great creators of modern science, Newton and Einstein, were driven by their creativity in science to an epistemological middle road. Newton chose a middle ground between Descartes and Bacon; Einstein between Kant and Mach. This middle ground is the sole avenue to reality and the sole secure foundation for scientific objectivity. It puts reality first, objectivity second, and method third. But precisely by putting reality first, the subjective scientist secures objective science by taking a position which is genuinely, though not superficially, subjective. This is what Polanyi unerringly sensed when he spoke of science as a personal knowledge.[39] He was too good a scientist to plead for shallow personalism in science. Had he been better informed about the epistemology of some "old metaphysics," he might have even saved himself from being bogged down in a "tacit knowledge,"[40] closely related to Gestalt psychology.

In a far deeper sense than the Gestaltists ever could dream, knowledge is personal. Knowledge is our personal, conscious commitment to a reality involving ourselves as well as that realm which exists independently of us. This personal commitment is so deep a form of subjectivity as to touch directly on the real and through the real on what is called the scientifically objective. Perhaps this is what was meant by that title which I found among the index cards on objectivity, a title which read: "deep subjectivity." I have not looked up the book itself, and I hope that placing that card there was not a librarian's oversight. For the title, "deep subjectivity," is the deepest that can be said about the real and the safest that can be said about the objective.

This is also what is implied in the three lectures given today on objectivity from the viewpoint of a mathematician, a physicist, and a neurobiologist. While a mathematician need not be even partially sane and can say whatever he likes within the limits of consistency, the mathematician cannot help wondering why of all his consistent systems only one fits the physical universe. What is so particularly "unreasonable" in this fact[41] from the empiricist viewpoint is that this particular system can be constructed without any specific reference to empirical reality. The "objective" truth of that system is, of course, established through experimental verification, but this "objectivity" already presupposes our ability to be in conscious touch with reality. Modern physics has amply demonstrated that the old mechanistic, Cartesian notion, in which method came first, objectivity second, and reality third, if it came at all, is untenable scientifically, a lesson which most biologists, to say nothing of sociologists and psychologists, are still to learn. They and we all should in addition realize that our work as thinkers is based on our use of the brain. What is being unfolded about its working can give no comfort either to the idealist or to the empiricist. Empirical sensations do not of themselves create ideas; such is at least a chief lesson of modern neurobiology. Ideas in turn do not create reality, or else everyone could be his own Einstein. What in particular we know about the consciousness of the brain is that even in its most abstract level of thinking it is never without a sensory ingredient which involves not only the thinking person but also his impersonal ambiance. Once this is acknowledged, the scientist, however subjective, will be aware of a reality which he does not create but only investigates with a method which retains its objectivity[42] only as long as there is no weakening on the part of the scientist that it is his faith swelling up from the very roots of his subjective, personal consciousness that puts him in touch with external reality. The particular results of this process can be verified, falsified,[43] corrected, improved, and qualified, but the process itself cannot be supplemented by anything else. It is the rock bottom of the objective work of scientists, however subjective, as we all are, even in such simple matters as the selection of the most relevant, that is, objective titles in a subject catalogue.

[1]The cards in question are those in the Firestone Library of Princeton University.
[2]Galileo Galilei, *Dialogue Concerning the Two Chief World Systems*, translated by S. Drake (Berkeley: University of California Press, 1962), p. 53.

Galileo wondered (ibid., p. 230) whether in the "whole ordinary philosophy" there was any such "rigorous proof" as his law of acceleration.

[3]The Symposium in question was the Colloque de l'Académie Internationale de Philosophie des Sciences held September 7-9, 1964. Its discussions were published under the auspices of *Dialectica* under the title, *Objectivité et réalité dans les différentes sciences* (Bruxelles: Office International de Librarie, 1966).

[4]Ibid., p. 241.

[5]For details, see Part I in E. Cassirer, *The Problem of Knowledge: Philosophy, Science, and History since Hegel,* translated by W. H. Woglom and C. W. Hendel with a Preface by C. W. Hendel (New Haven, Conn.: Yale University Press, 1950) and Chapter III in my *The Relevance of Physics* (Chicago: University of Chicago Press, 1966).

[6]B. Russell, "Recent Work on the Principles of Mathematics," *The International Monthly* 4 (1901), p. 84.

[7]Quoted in B. Jaffe, *Michelson and the Speed of Light* (Garden City, N.Y.: Doubleday, 1960), p. 93.

[8]I. Scheffler, *Science and Subjectivity* (Indianapolis: Bobbs-Merrill, 1967).

[9]See R. Batchelder, *The Irreversible Decision: 1939-1950* (Boston: Houghton Mifflin Company, 1962), p. 46, where it is also reported that Byrnes found with considerable relief that L. Szilard's strikingly pro-Soviet position concerning the arms race was not shared by other experts on atomic matters. Almost two decades later, following the memorable explosion by the Soviets of their huge, 60-megaton H-bomb in 1961, President Kennedy called in H. Bethe and E. Teller for consultation. After the separate interviews, in which Bethe argued against and Teller in favor of the resumption of atmospheric testing of nuclear bombs, Kennedy told J. Wiesner, his scientific advisor: "Well, I've done the right thing, I've listened to both of them, and they contradict each other beautifully. Now I'm free to do whatever I want." Reported by L. Edson, "Scientific Man for All Seasons," *The New York Times Magazine,* March 10, 1968, p. 127.

[10]A. Standen, *Science Is a Sacred Cow* (New York: E. P. Dutton, 1950).

[11]A. Koestler, *The Sleepwalkers: A History of Man's Changing Views of the Universe* (New York: Macmillan, 1959); F. A. Manuel, *Isaac Newton: A Biography* (Cambridge, Mass.: Harvard University Press, 1967).

[12]When this is not done judiciously, a disparagement of science is the result, which in turn can provoke scientists to look askance at historical studies of their fields, a process which is the topic of the well documented article of S. G. Brush, "Should the History of Science be Rated X?" *Science* 183 (1974), pp. 1164-1172.

[13]Fortunately, this fact is amply recognized and articulated by leading men of science. See, for instance, the Presidential Address of A. V. Hill at the Belfast meeting (1952) of the British Association for the Advancement of

Science, reprinted in A. V. Hill, *The Ethical Dilemma of Science and Other Writings* (New York: Rockefeller Institute Press, 1940), p. 84.

[14]Their convictions contrast sharply with the views of most philosopher-interpreters of science during those centuries, Malebranche, Berkeley, Locke, Hume, Kant, Schelling, Hegel, Comte, and Mill, a point which is the topic of several of my Gifford Lectures given at the University of Edinburgh in 1974-75. The series to be given in 1975-76 deals to a large extent with the question of objectivity in twentieth-century science. [Both published as *The Road of Science and the Ways to God* (Chicago: University of Chicago Press, 1978)].

[15]A point impressively documented as far as Einstein is concerned in Gerald Holton's "Mach, Einstein, and the Search for Reality," *Daedalus*, Spring 1968, pp. 636-673; reprinted in G. Holton, *Thematic Origins of Scientific Thought: Kepler to Einstein* (Cambridge, Mass.: Harvard University Press, 1973), pp. 219-259.

[16]Of these, Bohr and Born were on the positivist side with their endorsement or rather formulation of the Copenhagen interpretation of quantum mechanics. Schrödinger, and to a lesser extent De Broglie and Wigner, sympathized with Einstein's position. More recently Heisenberg sought to reach a more metaphysical level though not too convincingly. Dirac remained conspicuously reticent on philosophical matters.

[17]According to a first-hand report of W. E. Hocking. See his *Science and the Idea of God* (Chapel Hill, N.C.: The University of North Carolina Press, 1944), p. 96.

[18]In a letter of Dec. 9, 1935, to A. Lampa. Einstein disputed the claim that Mach had fallen into oblivion. On the contrary, Einstein claimed, the philosophical orientation of physicists was very close to that of Mach. See Holton, art. cit., p. 668.

[19]In a less pointed manner such was the gist of E. Gilson's remark: "Nothing equals the ignorance of modern philosophers in matters of science, except the ignorance of modern scientists in matters of philosophy." See his lecture, "Science, Philosophy, and Religious Wisdom" (1952), in A. C. Pegis (ed.), *A Gilson Reader* (Garden City, N.Y.: Doubleday, 1957), p. 217. With less philosophical articulation, but with greater vehemence, A. Lande, who made important contributions to quantum mechanics, described as "magic mongering" the facile attitude with which most physicists ignored the ontological problem implied in their labeling the same phenomenon now as a wave, now as a particle. See A. Lande, *New Foundations of Quantum Mechanics* (Cambridge: University Press, 1965), p. 8.

[20]Among the others was E. Nagel, best known for his book, *The Structure of Science: Problems in the Logic of Scientific Explanation* (New York: Harcourt, Brace and World, Inc., 1961), but like Scheffler, he too wanted to vindicate objectivity on an essentially empiricist basis. Scheffler tried to

save objectivity by endorsing Reichenbach's distinction between the genesis of a discovery and its empirical justification. In the former, ideas (and metaphysics), even wholly arbitrary ones, were admitted, but not in the latter, which alone had rigorous connection with reality and objectivity (*Science and Subjectivity*, p. 73). Such a reality and objectivity were not, however, essentially different from mere sensations.

[21]See T. S. Kuhn, *The Structure of Scientific Revolutions* (Chicago: University of Chicago Press, 1962), p. 109.

[22]As remarked by Kuhn himself in the Preface of his book.

[23]In B. Russell's "Reply to Criticisms," in *The Philosophy of Bertrand Russell*, edited by P. A. Schilpp (3d ed.; New York: Tudor Publishing Company, 1951), p. 697. Typically, Russell's remark is part of his comments on Einstein's contribution, "Remarks on Bertrand Russell's Theory of Knowledge," which comes to a close with a comment on Russell's *An Inquiry into Meaning and Truth* (1940). Since in that book Russell markedly shifted toward metaphysics, Einstein noted: "The only thing to which I take exception there is the bad intellectual conscience which shines through between the lines," ibid., p. 291.

[24]See note 20.

[25]Sartre's sharp remarks about science are perfectly logical in view of his basic perspective which cares only for the moment, whereas for science that aspect of reality is of paramount interest which transcends the moment and is therefore predictable and subject to anticipated control.

[26]These subsequent centuries are characterized by the striking contrast between the ideals of political harmony set forth by the great Greek philosophers and the continual dissension among the Greek city-states.

[27]Hesiod's account has some similarity with the theogonical cosmogony of the *Enumah Elish*. Both presuppose nature to be animated and therefore subjective to a large extent. Quite different is the world-view of biblical cosmogony in spite of its Babylonian framework. In the biblical account nature works objectively because it is the product of a non-capricious, fully rational, personal Creator. See on this my *Science and Creation: From Eternal Cycles to an Oscillating Universe* (Edinburgh: Scottish Academic Press, 1974), Chapters V-VII.

[28]In Chapter I of my *The Relevance of Physics* I tried to do justice to the fact that Socrates was the proponent, and a most influential one, of a new physics, a point largely ignored in histories of science in spite of Leibniz's repeated references to the importance of *Phaedo* for physical theory. For details, see *Leibniz Selections*, edited by P. P. Wiener (New York: Charles Scribner's Sons, 1951), pp. 69, 320, 322, and 325.

[29]"Wretched mind," Galen lets the senses rise in indignation against the speculative atomist admitting only extension and motion, "do you, who got your evidence from us, yet try to overthrow us? Our overthrow will be your

downfall" (Diels 68B 125). The profound truth of this remark should be clear to anyone familiar with the effectiveness with which Berkeley used the arguments against the reality of secondary qualities to cast doubt on the reality of the primary qualities.

[30]Through star-gazing, scientific or otherwise, Hume declared in the *Treatise of Human Nature* (Book I, Part II, Section 6), "we never really advance a step beyond ourselves, nor can we conceive any kind of existence, but those perceptions, which have appear'd in that narrow compass." Consequently, man remains forever confined, rigorously speaking, to the "universe of imagination," and he can have ideas only of "what is there produc'd." In stating this Hume was perfectly consistent with his premises, but it should also be clear that when those premises are taken seriously and consistently, scientific work becomes meaningless. It is this consistency which is invariably ignored by scientists glibly endorsing Hume's philosophy or some of its equivalents. One such world-renowned astronomer, who insisted at an international conference on cosmology that looking through a telescope merely changes our sensations, answered with a surprised "I am really not sure" to my question whether in his view the wall facing him was merely a sensation of his, or existed independently of it.

[31]See his paper, "On the Utility of Speculative Inquiries," read on April 14, 1780, before the Philosophical Society of Bath, in *The Scientific Papers of Sir William Herschel* edited by J. L. E. Dreyer (London: The Royal Society, 1912), vol. I, pp. xxxi-xxxiii Herschel criticized a member of the Society who claimed in genuinely Humean style that all scientific discourse should be restricted to matters of fact, namely, sensations.

[32]In a letter of January 6, 1948, to Michele Besso. See the *Einstein-Besso Correspondence,* translated into French with notes and introduction by Pierre Speziale (Paris: Hermann, 1972), p. 391.

[33]This is what underlies Einstein's appraisal of the philosophical attitude of most physicists as being equivalent to playing, unwittingly, a risky game with reality. See his letter of December 22, 1950, to Schrödinger, in *Letters on Wave Mechanics,* edited by K. Przibram, translated by Martin J. Klein (New York: Philosophical Library, 1967), p. 39.

[34]B. Russell, *An Inquiry into Meaning and Truth* (London: Allen and Unwin, 1940), pp. 148-149.

[35]B. Russell, *Philosophy* [in the English edition, *An Outline of Philosophy*] (New York: W. W. Norton, 1927), p. 14. In speaking of the enigma of induction Russell stated there that "until it is solved, the rational man will doubt whether his food will nourish him, and whether the sun will rise tomorrow. I am not a rational man in this sense, but for the moment I shall pretend to be." Needless to say, Russell expected the solution to be had on an empir-

ical basis, a patent impossibility because all induction is steeped in meta-physics.

[36]Ibid., p. 80.

[37]This is the epistemological creed of moderate realism, in substance as old as the median position which Aristotle tried to steer between Plato and Heraclitus. The conviction that this intimate union of the mind with exter-nal reality is the primary datum of epistemology is the gist of the three steps on the road to objective reality as specified by E. Gilson in his *Le réalisme méthodique* (Paris: Téqui, 1938), p. 87. The first step is the recog-nition that one has always been a realist; the second is the admission that it is impossible to think but in a realist way, the third is that critics of realism also think in a realist way as soon as they forget their assumed posture. Once such critics ask why is this so, Gilson concludes, "their conversion is almost complete."

[38]When one starts, as Descartes did, from the abstract mind, external real-ity becomes an illusion in the long run at least. From the opposite starting point, the realm of pure sensations, one cannot even reach, as Hume had to admit, one's own personal identity. As to Kant, it was aptly remarked: "He started from both ends of the road at once, but he never met himself." Nor did he meet, one may add, external reality, or the *Ding an sich*. The fore-going remark is attributed to Edward Craig by W. Temple in his *Nature, Man and God* (London: Macmillan, 1934), p. 70.

[39]M. Polanyi, *Personal Knowledge: Towards a Post-Critical Philosophy* (Chicago: University of Chicago Press, 1958). For an excellent discussion of Polanyi's epistemology, see the essay by T. F. Torrance, "The Place of Michael Polanyi in the Modern Philosophy of Science," where Polanyi's views are discussed with an eye on the thought of Einstein, Popper, Bohr, and Gödel.

[40]This is why Polanyi could become the easy, though basically unjustified target, of Scheffler, who saw no difference between Polanyi's and Kuhn's positions. See Scheffler, *Science and Subjectivity*, p. 74.

[41]This fact is the topic of E. P. Wigner's remarkable article, "The Un-reasonable Effectiveness of Mathematics in the Natural Sciences," *Com-munications on Pure and Applied Mathematics* 13 (1960), pp. 1-14.

[42]It is of no small epistemological significance that the fact and practice of exact science fail to be derailed from the tracks of objective reality while modern epistemological trends are careening in ever wilder swings.

[43]In view of the prominence achieved by Popper's theory of scientific knowledge as being a sequence of falsifying ever broader propositions, a brief remark will not be out of place on its foundation, or rather on the lack of it. Popper seems to overlook the fact that even the most primitive "empirical" statements imply generalizations which are always a form of induction, or a kind of metaphysical assertion of objective reality. For all

his declarations about being a realist, Popper seems to be ready to march on the road of realism without the willingness to make the first step in an explicit manner. But unless we have verisimilitude (a notion on which, in a truly realist style, Popper sets so great a store) between concepts and facts at the very start, no further improvements on that verisimilitude can be made with subsequent falsifications. For Popper's emphasis on verisimilitude and on his self-identification as a realist, see his *Objective Knowledge: An Evolutionary Approach* (Oxford: Clarendon Press), p. 318 as a classic locus.

3

Maritain and Science

"Few spectacles are as beautiful and moving for the mind as that of physics thus advancing toward its destiny like a huge throbbing ship." Such are the words that introduce Maritain's discussion of "The New Physics."[1] Almost exactly half-a-century old,[2] that discussion should for that reason too be worth considering at this celebration of the centenary of Maritain's birth. Praising the new physics was by 1932 a useful foil for philosophers, and for physicists waxing philosophical, to treat with a pretense to originality of difficult topics without saying anything profound about them. Space, time, causality, external reality, motion and matter—so many crucial issues for physics—have always been difficult topics for

---

This paper was presented at the meeting which the American Maritain Association held at Princeton University, October 28-29, 1983, in commemoration of the centenary of Maritain's birth. Since shortly after that meeting I was able to consult the archives of Lycée Henri IV and of the Sorbonne, and relevant dossiers in the Archives Nationales, I felt it appropriate to rewrite and expand the section dealing with Maritain's student years. Reprinted with permission from *The New Scholasticism* 58 (1984), pp. 267-92.

philosophers who were scrupulously exacting of themselves. Such a philosopher was Jacques Maritain. According to his own admission he dwelt at some length on the new physics "not to indulge in any rash prophecies on the future of its theories, but to see whether its scientific progress confirms or invalidates the epistemological principles we have been trying to establish up to this point."[3]

That a serious metaphysician ought to be seriously concerned with physics in particular or with knowledge about sensory nature in general, should seem obvious.[4] Such concern is almost like a somber warning to the metaphysicians of modern times. Science, especially physics, has increasingly taken over the center stage of intellectual interest for the past 300 or 400 years. To be sure, the intellectual talent needed to cultivate metaphysics is not exactly the same as the bent of mind needed in the exact sciences. There is, however, no doubt that during the last 300 or so years science kept drawing to itself talented minds who in former times most likely would have ended up as metaphysicians if not plain theologians.[5] Whatever one may think of the intrinsic merits of Hegel's philosophy, which was all metaphysics, he displayed extraordinary mental powers. A man with similar powers has not since appeared on the metaphysical scene. Tellingly, Hegel finished his career when Faraday started his with the discovery of electromagnetic induction. It was a discovery which in a sense was more instrumental in ushering in the new physics, that is, electromagnetic theory, relativity, and quantum mechanics, than were all the scientific discoveries prior to 1832, including the discoveries of Newton and Galileo.

From the mid-nineteenth century on scientists began to feel an undisputed sense of superiority over metaphysicians. Of course, the metaphysics they mostly knew was the metaphysics of German idealism[6] which precisely because it presented itself with so superior an air, could readily appear in its very lowly nature. It certainly deserved the sarcasm of Maxwell who around 1870 spoke "of the den of metaphysicians, strewed with the remains of former explorers, and abhorred by every man of science."[7] In the early 1930s, just about the time when Maritain's reflections on the new physics saw print, no less a prominent physicist than Max Born brushed aside what he called the "dry tracts of metaphysics."[8] Still another half a century later, the physicists' dislike of metaphysics is as strong as ever. That there is no metaphysics whatsoever in physics was the flat dismissal of a set of questions which I posed two years ago, I believe, to a world-renowned physicist well known for his work on scientific cosmology. Strangely enough the same physicist failed to remember that he hardly ever fails to cite with approval Bishop Berkeley's idealist metaphysics, the logical outcome of which is solipsism, the supreme form of misguided metaphysics.[9] It would be mistaken to see a bellwether in the often-quoted statements with which Einstein, from his fifties on, endorsed realism and metaphysics. However enlightening, those statements impressed at most only a few philosophers of science but hardly the body scientific.[10]

Such is the general background against which Maritain's appraisal of the new physics should be seen. It is not an appraisal which, however just and incisive, would turn the thoroughly anti-realist and therefore thoroughly antimetaphysicist tide which flows back and forth over the entire intellectual landscape of our times. Such tides are fashions against which, just as against huge tidal waves, no argument, however rational, has ever proved convincing. Maritain, it is well to recall, was not only fully aware of this situation but also described it in phrases characteristic of his unsparing incisiveness.

The particular background against which Maritain's dicta on the new physics should be seen concerns Maritain himself. While a great deal is known about most aspects and phases of Maritain's career, few specifics are available in the printed record about Maritain's two years at the Lycée Henri IV, one of France's prestigious lycées, and about his six years at the Sorbonne, 1900-1906. The primary published source on those years has been Raissa's account of her and Jacques' spiritual odyssey.[11] There we are told about the precocious philosophical anxieties which beset Jacques as a student at Henri IV[12] and about the good prospects which he had at the Sorbonne owing to the esteem in which he and Raissa were held by Felix Le Dantec, their teacher of zoology.[13] Raissa also recalled the names of teachers whose courses in the life and earth sciences they followed at the Sorbonne.[14]

Had Jacques ambitioned a study of the exact sciences (mathematics and physics) at the Sorbonne, he would have spent two more years at Henri IV, the years known as *mathématiques élémentaires* and *mathématiques spéciales*. The two years preceding these were called *rhétorique* and *philosophie* which a typical student attended in his sixteenth and seventeenth years. In both of those years the classes were the ones prescribed by the ministerial decree of February 14, 1885, which set the pattern, a rather rigid one, of French secondary education for decades to come.[15] In the *rhétorique* the courses included algebra with trigonometry and a mostly descriptive physics with astronomy. The registry of examinations in Henri IV contains only the marks which Jacques obtained in Latin, French, German, and history.[16] They reveal a student who impressed his teachers not so much with his actual performance as by his inquisitive mind. According to his French teacher he was "very intelligent, somewhat distracted, not the student type, one who could do better by not thinking also of matters other than those discussed in class."[17] His Latin teacher described him as an "esprit net, de la finesse," no small praise coming as it did from Georges Felix Edet, a renowned classicist.[18] For his history teacher Jacques was "intelligent, un peu inégal." Only the German teacher found his scholastic achievement above any criticism.

As to the *philosophie* year the entries in the registry are very incomplete concerning Jacques. One finds there only the marks which he received in philosophy in all three trimesters and the mark given him in physics and chemistry in the third trimester. For the philosophy teacher, Hector Dereux, Jacques was an "excellent student, with an adventurous though fine and distinguished mind, a searching spirit worthy indeed of esteem and sympathy."[19] The intellectual profile of Maritain is clearly forecast in these observations. Jacques had to take a course in natural history as well, in addition to German, history, and a review course of algebra and trigonometry (calculus was reserved for *mathématiques élémentaires* and *spéciales*). The *Annuaire* of Henri IV for 1899-1900 reveals that Jacques took first prize in natural history.[20] Clearly, he was well prepared to matriculate in the Sorbonne either in the Faculté des Lettres or in the Faculté des Sciences. He entered both in November 1900. On July 26, 1902 he obtained his *licence* in philosophy in the Faculté des Lettres.[21] Obtaining *licence* in the life sciences was a longer affair, partly because of the laboratory work involved.

For *licence ès sciences* one had to obtain three certificates in any of the three major fields—mathematics, physics (chemistry), and biology (geology), in each of which, especially in the latter, a fair choice of subjects was available. Jacques obtained a certificate in botany in July 1901, one in physiology in July 1903, and one in mineralogy in July 1906. His personal card[22] also contains references to his attending a course in geology in 1902-03. Of course, he may have audited other courses as well, and, as will be clear shortly, he must have attended courses in evolution and embryology. His marks were *bien* and *assez bien* which put him in the top fifth. In botany his teachers were Prof. Gaston Bonnier (1853-1922) and his associate, Louis Matruchot (1863-1921). The physiology courses were taught by Prof. Jules Dastre (1844-1917) and his assistant, Louis Lapicque (1866-1948), who became professor in 1919. Mineralogy was taught by Prof. Emile Haug (1861-1927) and his assistant, Louis Gentil (1868-1925).

Such was a distinguished group of scientists most of whom have received a special article in the *Dictionary of Scientific Biography,* in witness of the lasting value of their contributions. The same is true of Felix Le Dantec (1869-1917), assistant during Jacques' Sorbonne years to Alfred Giard (1846-1908), professor of zoology, both of whom taught courses on evolution as well. Le Dantec, who became professor of biology in 1908, taught also a course in general embryology. A widely read popularizer as well, with a vibrant and warm personality, Le Dantec made quite an impression on both Jacques and Raissa, who obviously felt that Le Dantec had a great and influential scientific career ahead of him. Such is the background of Raissa's remark that Le Dantec held out for Jacques the prospect of a brilliant scientific future. But then Jacques would have had to commit his life to Le Dantec's chief aim, namely, producing living and conscious matter in vitro.[23] While Le Dantec was an outspoken materialist,[24] his two favorite students had by then got beyond that stage of intellectual perception in which eating is simply taken for proper nourishment. By 1906 both Jacques and Raissa had suffered an intellectual hunger which drove them at one point almost to suicide.[25] Their first move to satisfy their hunger consisted in embracing a study of biology which plainly acknowledged the insufficiency of physics to deal with the immense richness and adaptability of organic life. The academic years 1906-07 and 1907-08 found Jacques and Raissa in Heidelberg where Jacques studied with Driesch, a leading German biologist of the time and the head of the neovitalist movement.[26] Jacques became the first in 1910 to introduce Driesch's work and thought to the French intellectual world, which was not, however, to pay attention to an article published in a Catholic philosophical monthly.[27] Ten years later he wrote an introduction to the French translation of Driesch's magnum opus on the science and philosophy of organism,[28] first delivered as Gifford Lectures at the University of Aberdeen in 1907 and 1908.[29]

Maritain's article published in 1910 was his first and last article that could be considered as a scientific paper. Quite possibly, if the scholarship that took him to Heidelberg had not included an obligation to present in a publication his studies there, that article might not have been written. Actually the two years which separated Maritain's return from Heidelberg[30] and the publication of that article, a mere 25 pages, suggests that Maritain's heart was no longer in the sciences. Science as a professional study ceased to exist for Maritain at almost the very moment he had completed that year in Heidelberg. Moreover, what Anatole France (that is, Jacques Thiébault) had by then described with a literary finesse, which was persuasive in the measure of its professed cool detachment, almost happened to Maritain. France spoke of his own love turned into hatred as he made one of his heroes, Jerome Coignard, state: "I hate science for having loved it too much after the manner of voluptuaries who reproach women with not having come up to the dream they formed of them. I wanted to know [through science] everything and I suffer today for my culpable folly."[31] Indeed that former lover of science was to speak of it in terms of vibrating contempt: "The universe which science reveals to us is a dispiriting monotony. All the suns are drops of fire and all the planets drops of mud."[32]

Maritain never despised science, not even under the impact of transferring his intellectual loyalties to metaphysics in general and to Thomism in particular. Nothing would have been more natural for him than to follow, say a Brunetière, certainly a leading figure around 1900 on the French intellectual scene, whose phrase, "the bankruptcy of science"[33] had become a byword and a battle cry by the time Maritain entered the Sorbonne. While even Brunetière failed to see the difference between science and scientism, young Maritain unerringly put his finger on scientism as the true cultural culprit and curse.[34] He did so in an article published in June 1910, which was in fact his first publication and carried the title: "La science moderne et la raison" (Modern Science and Reason),[35] an article still to appear in English translation. Some of the major themes of what Maritain was to say later on the subject in his most mature works are fully stated in that article. Unlike those works, that article vibrates with the unabashed fiery élan of the one who at the age of 27, a rather youthful age, finds the long-sought answer and who is suddenly seized with a missionary consciousness to share his find with his fellow men.

That Maritain was already at the age of 27 the kind of thinker he wanted to be remembered, and who is in fact remembered as such, has a partial explanation in a reminiscence of Raissa about their Sorbonne years. What would have happened, Raissa asks, if "we had directed our studies toward the physico-mathematical rather than toward the natural sciences," that is, botany, physiology, embryology, and geology? It would have been a very mixed blessing. On the one hand, "We would undoubtedly have been fascinated by the magnificence of the discoveries of so great a galaxy of scientific genius; . . . it would have been wonderful, for instance, to attend the classes of Paul Appell, or of Marie and Pierre Curie, scientists of genius and heroic workers, who had opened the way to a new science." On the other hand, fascination by what was presented about the latest marvels of physical science "would for a long time have hidden from us our hunger for metaphysical knowledge."[36] The gist of this remark is that the exactness of physical science is more effective than is the partially descriptive character of other empirical sciences in distracting one from questions about the causes, essence, and purpose of empirical reality. This is not to suggest that professors in the Faculty or rather departments of natural sciences at the Sorbonne cared for such questions. When Raissa posed such questions to Prof. Lapicque, he brushed her aside with the words: "But that is mysticism."[37]

Mysticism was to become a legitimate and integral subject matter in the metaphysics of Maritain. The last third of his *Degrees of Knowledge* is about the contemplation of the highest truth, God. To try to answer questions about cause, essence, and purpose has always been an enterprise which inevitably conjures up questions about the ultimate cause, the ultimate essence, and the ultimate purpose. The philosopher may, of course, rush to the ultimate. Indeed, as they try to gain hold of the ultimate many philosophers jump one or two if not all the intermediate degrees of knowledge. Descartes was one such philosopher. Convinced that God's nature was an open book to him, Descartes thought, and very logically, that he had therefore a privileged access to the entire observable nature. The result was not a science about the universe of things, but a mere novel, to recall the biting remark of Huygens, a younger contemporary of Descartes and of incomparably greater stature as a physicist than Descartes was. In the case of Hegel the result was the *Enzyklopädie der philosophischen Wissenschaften,* a horror-story about science.[38] The other side of the coin is no less instructive. Philosophers, whose principal reasoning was aimed at establishing man's inability to know the ultimate or God, fared just as badly with respect to their legislation about science. The dicta and at times lengthy discourses of a Hume, a Kant, a Mill, a Spencer, a Mach, and of a Carnap on how science should be cultivated amount to a scheme of how to put science into a strait-jacket.[39]

With both extremes, the rationalist and the empiricist, the source of the fiasco lies with the concept of reality. In their discourse on reality the modern rationalists as well as the modern empiricists were heavily conditioned by what they thought to be the doctrine of the science of their day about reality. They all failed to see the difference between the science of their day and the prevailing fashionable views about it. Those views were not science but philosophy or metaphysics, and all too often a very corrupted form of metaphysics. Maritain never made that mistaken identification. In fact, in the page which in a sense is the first page of his *Degrees of Knowledge,* he begins with a clear-cut distinction between science and philosophy, that is, a critico-realist philosophy. He did not make that distinction in order to get science out of the way. On the contrary, he was eager to show that whereas the positivist scheme does not at all accord with science exactly, the critico-realist philosophy "corresponds more exactly to the vast logical universe of whose modern development the sciences offer us some picture."[40] Maritain added in the same breath that a setting forth of their correspondence would require an entire book, a book which he described as a sort of history of science. Maritain never wrote that book, a fact about which the historian and philosopher of science, such as myself, can only feel a deep frustration. Not that around 1920, let alone earlier, the writing of such a book would have been a relatively easy task except for its pre-Galilean part which had received in Duhem a brilliant pioneer interpreter with whose work Maritain was fairly familiar.[41] Histories of science available until recently were top-heavy on clichés forged by the Enlightenment and by its heir, positivism. Most of those histories were conspicuously void of positive facts. While such an imbalance could and did in fact mislead most readers, Maritain would have unerringly seen the difference between the grain and the chaff.

This last appraisal would easily pass for hero worship were it not for the evidence that Maritain's reading of the literature on the new physics was very judicious. It was a literature in the sense that Maritain in all evidence never read strictly scientific publications relating to the new physics. His scientific education at the Sorbonne was centered on the biological sciences. Whatever free time he had, he spent it, not by reading calculus and rational mechanics, but by attending Bergson's lectures and by being a student activist. In fact, he met Raissa as he gathered signatures for a protest against Czarist pogroms. In writing about the new physics Maritain had to rely on what the French felicitously call "haute popularisation." They included books and lectures by British, German, and French men of science.[42] Among the British were Eddington, Whitehead, Russell, W. R. Thompson; among the Germans, Einstein, Weyl, Herzfeld, Heisenberg; among the French, Picard, Langevin, Poincaré, Urbain, De Broglie. Mention should also be made of first-rate presentations of the philosophy of physics and mathematics by Meyerson, Duhem, and Gonseth. The latter's book on the foundations of mathematics and the four-dimensional space-time manifold[43] was on a level even higher than the very highest popularization. Last but not least mention should be made of Maritain's personal contacts with scientists such as his conversation with Lebedeeff,[44] the first to measure the pressure of light, and of Maritain's close following of the lectures which Einstein gave in Paris in April 1922. More of that later. The footnotes of the *Degrees of Knowledge* also contain references to several books and articles published in 1930 and 1931, in proof of Maritain's resolve to remain abreast as much as possible with the latest about the new physics.

The new physics, it is well to recall, was around 1930 a mere dwarf compared with what it was to become in the following half a century. To give only two examples: nothing was known in 1930 about the antimatter which now plays a central role in theories relating to the earliest phases of the universe. The first evidence of antimatter was discovered only in 1932. Again, nothing was known yet about the neutrino, first postulated in 1934 by Fermi, which today is the hottest topic in fundamental particle physics. Today when atoms, to say nothing of molecules, can be directly observed through electron microscopes, and when atomic energy has entered countless households as source of electricity, it is difficult to recall the time, only 50 years ago, when one could say as Maritain did that "atoms are symbolic images of the primordial parts of the spatio-temporal organization of matter." Such was his answer to the question as to *what is* (*quid sit*) an atom or an electron. Not that he doubted their reality. "They appear," he wrote in answer to the question whether they exist (*an sit*) "to be realities (something exists which the words electron and atom circumscribe determinately)."[45] Maritain's distinction between the answers given to *an sit* and *quid sit* is still called for concerning particles constituting the atomic nucleus and their subconstituents. Maritain expected precisely that type of development illustrated by the ever receding fundamental level among elementary particles whose never-ending proliferation could in 1932 appear but a dream.

Maritain's expectation was based on what he called "the real being as an inexhaustible source of effectuable measurements,"[46] that is, a subject which permits an ever more detailed mathematicization. With this we have arrived at the central problem of a critico-realist philosophy with respect to science, be it the new physics. The problem is the extent to which the mathematical models built by the theoretical physicist correspond to reality. Such is also a problem for the theoretical physicist. Yet the respective concerns of the realist philosopher and of the theoretical physicist have a different dynamics. The difference relates to the primary meaning they respectively attach to the term real. The difference, as Maritain put it, in a memorable paragraph which renders graphically the respective thinking of the two, is a matter of emphasis. The philosopher must put the emphasis on the difference between a real being and a mere being of reason, such as a mathematical function. The latter is, however, for the physicist the real thing provided it works. "Doubtless replies the philosopher. That is what they are made for, even those among them which are most obviously *entia rationis*. And, not to be outdone by his colleague, the physicist will immediately add that these real entities are 'shadows' or allusions from which it would be silly to expect anything concerning the intimate nature of matter."[47] Such a graphic hint of the unity of the two viewpoints is a priceless example of the

gist of Maritain's *Degrees of Knowledge,* a title which in its full form
begins: "Distinguish to Unite." Of course this full title made sense
only if there was an ontological ground for unity. Not that Maritain
would have ever made that unity depend on an opinion poll, be it
the polling of leading physicists. But he knew all too well that in a
scientific age the words of scientists carried a special weight. He
was among the first to see the significance of Einstein's increasingly numerous votes on behalf of ontological reality as presup-
posed by physical theory.[48]

The relation between reality and mathematical symbolism used
in physics was only one of the great problems which imposed itself
with a renewed force on the philosophical consciousness. Another
was the true sense of the relativity of time and motion, in other
words, the possibility of simultaneity. Curiously, not one among the
many physicists and philosophers of science, who discussed this
question since the 1930s, cared to seize upon an astute remark
of Maritain which first appeared in the wake of Einstein's lectures
at the Sorbonne in 1922. The remark concerned Einstein's continual return to the question "What does the simultaneity mean for
me the physicist?" He always replied, Maritain recalled, by falling
back on the methodological theme: "Give me a definition that will
tell me by what ensemble of measurements, concretely realizable in
each case, I can verify that two events deserve to be called simultaneous or not; only then will I have a definition of simultaneity
which can be handled by a physicist and have value for him."[49] In
other words, Einstein implicitly acknowledged the irrelevance of
the empiriometrical method to ontological questions about simultaneity and even about the speed of light. Ontological questions had
therefore to be referred to that much misunderstood and much
abused common sense with which fewer and fewer intellectuals
were willing to appear in public. Physics, especially the Einsteinian
physics of relativity, may have liberated itself completely of common sense, Maritain remarked, but only at a price. The bargain
meant a complete renunciation of any assertion about the real, a
renunciation to which no physicist could consistently subscribe.
Furthermore, the complete relativization of empiriometric procedures depended on keeping in the same form all basic equations
of physics, regardless of their reference system. What this meant
was an absolutization far more profound than, say, the postulating
of an absolute space in Newtonian physics.[50] Very few physicists

and philosophers of science saw around 1930 Einstein's physics in that absolutist light.[51] Maritain's insight was not acknowledged by those still not too many who in recent years have spoken and written as if they were the "discoverers" of that absolutist aspect. Even today a mere mention of that aspect often leaves dumbfounded the scientific and intellectual community thriving as ever on hollow clichés as to what Einstein's relativity is really about.

Another such cliché covering a major problem raised by the new physics is the question of causality. That quantum mechanics in general and the uncertainty principle in particular have given the coup de grâce to causality is an assertion which saw print in countless contexts. It has in fact become a climate of opinion ever since that principle was formulated by Heisenberg who rushed to give it that anticausal interpretation.[52] It shows something of the philosophical poverty of the scientific world that it fails to realize the obvious, namely, that empiricism, be it mathematico-physical in the most refined form, is incapacitated by its professed empiricism to say anything about causality. While empiricists are not different from anyone in that they can *observe* only sequences, empiricists as such can *know* only mere sequences. "Causation," Maritain recalled the elementary truth, "is not *observable* as such or insofar as it is an intelligible relation."[53] Moreover, as Maritain also pointed out another elementary truth, failure to measure an interaction exactly in an operational sense could not mean that the interaction could not take place exactly in an ontological sense.

There were some well-meaning physicists, Heisenberg himself and Eddington in particular, who tried to undo their destruction of causality with the claim that the freedom of the will might therefore be rehabilitated in a world which is non-causal. Maritain (we are in 1931 or only four years after the enunciation by Heisenberg of the principle of indeterminacy) tersely noted that not only does the principle have no bearing on the freedom of the will, but it has no bearing whatever on purely physical causality.[54] In substance, Maritain characterized as nonsense any abolition of causality and any reinstatement of freedom on the basis of the uncertainty principle. He did so four years before Eddington, who had even tried to calculate the physical measure of freedom allotted to the human will by the new physics, came to the conclusion that such a calculation was sheer nonsense.[55]

Time does not allow us to review and appraise, however briefly,

Maritain's analysis of what the new physics offered by 1932 on space, multidimensions, and space-time manifold.[56] Nor does time allow us to review his strictures of those physicists who in the name of quantum mechanics had abolished individual reality.[57] It should, however, be pointed out that Maritain did not spare from criticism those of his fellow Thomists who were swayed by some apparent similarities between the ideology of the new physics and several basic doctrines of Aristotelian realism.[58] One of these is hylomorphism, according to which actual reality is the actualisation of a purely potential entity called *materia prima*, by a so-called substantial form. Just as the latter is not the observable shape, the former is not a matter which would have any, even indirectly observable features. In defense of that often vilified doctrine of hylomorphism, let me recall only the fact that any philosophical school which tried to cope in terms of another doctrine with the problem of identity through change, either had to deny change or had to throw consistency and coherence to the winds, or to pretend that change represents no problem at all. But let me return to the *materia prima* to which Aristotelians of all times have been tempted to give some observable properties. The temptation was either that of some misguided realism or, what is worse, the temptation to be in agreement with the latest intellectual or scientific fashions. Thus it happened that in the 1920s several prominent scholastic philosophers began to talk about the Bohr atom as an illustration of hylomorphism. The nucleus of the atom was taken for the *materia prima* and the arrangement of the electrons around it for the substantial form. Maritain had no use for a policy which by marrying the fashions of its own age is bound to become widowed in the one to follow.

In those intellectual fashions Maritain saw topics for the sociology of knowledge, a point that should seem quite original in view of the rather recent trend to see in all intellectual activity, including scientific activity, a mere matter for sociology. Not only original but also far more profound. This should be very clear in view of the latest phase of the sociologization of the analysis of scientific thinking. Its proponents are interested only in ever varying social patterns, or to use the prevailing jargon, in revolutions and mutations.[59] The utter superficiality and, I should add, irresponsibility of this allegedly scholarly interest become all too evident when it comes to the imperative of seeing connections among those successive patterns or paradigms. The imperative is imposed by the widespread belief in scientific progress. Nothing shows better the superficiality of paradigmists than their failure to show evidence of progress in the succession of paradigms. The root of their fiasco is their extracting philosophy from the new physics without seeing that in fact they exploit not the new physics or the new biology but the pseudophilosophy grafted onto it. This would not have happened if the sociology of science had observed a judicious remark of Maritain: "From the epistemo-sociological point of view science is no longer considered *in itself*, that is, in respect to what is true or what is false, nor in respect to the determinations which necessarily result from the exigencies of science in the knowledge of things. Rather, it interests us as a collective attitude or spirit engendered *hic et nunc* in the mind of men. So considered, it influences the evolution of the mind as a ferment or a center of organization of various activities belonging to an associative, rather than a rational order and accidental in relation to the essence of science itself."[60]

The gist of this remark is that a climate of opinion engendered by the science of the day was not philosophy, let alone a valid and good philosophy. Consequently, change in such a climate was not to be considered an obvious progress, an unmitigated blessing. To be sure, as Maritain observed, the ideology surrounding the new physics certainly discredited the mechanistic ideology grafted on classical physics. He viewed that outcome "a considerable result from the viewpoint of the sociology of the intellect."[61] But, as he added prophetically: "the new physics will influence the common intellect in the same irrational fashion as classical physics did. Through some sort of associative influence or sub-intellectual induction, it will probably give birth in its turn to an inchoate philosophy, a new 'scientific tableau of the cosmos' which will save us from the former errors only at the price of illusions of another type."[62] As a chief example of those new illusions in the offing Maritain singled out the saving of freedom by quantum mechanics through its destruction of causality. Then he turned to his other major topics of the analysis of modern science, namely, the respective roles of ontology and empiriology in the study of living organism.

Those respective roles, as one can easiy guess, related to the questions of what is life or living matter, and what is the purposiveness apparent in such a matter. In this very year, when the 25th anniversary of the discovery of the double-helix structure of DNA is being celebrated and heavy investments are attracted by the gene-splicing industry, it may sound almost irrational to hint that anything but sheer physics is involved in living matter, including its conscious kind. Yet it is well to recall that a very different kind of irrationality was seen by Sir Andrew Huxley to encroach upon biological research when three years ago he warned the scientific community against taking the origin of life as solved and against shoving the problem of consciousness under the rug, a still greater problem for biology.[63] Indeed if the charge of irrationality is to be brought against any trend, it is the trend according to which the purely physico-chemical account of life has for some time been completed. Anyone familiar with the world of biology could easily do today what was done in 1932 by Maritain who cited quite a few biologists who, although authorities in the latest phase of research, expressed their conviction that life is more than mere physico-chemical mechanism.[64] The survival for the past 150 years within biology of a prominent dissenting minority is in fact a most unique phenomenon in modern science, which witnesses the opposition party die out within a generation or so.

This opposition party in biology is usually referred to as the party of vitalists or neovitalists, terms rather misleading because they seem to imply that some sort of mysterious vital fluid is at work in living organisms. The vitalism Maritain defended was never of this sort. He insisted on the validity of a never-ending extension of the physico-chemical method[65] and renounced a vitalism which would thrive on an observable area denied to that method.[66] Maritain wanted no part in more subtle forms of vitalism either, such as the one proposed by Bergson,[67] or the one which is generated by phenomenology.[68] The latter was especially dangerous in Maritain's eyes because of its professed emphasis on doing justice to all phenomena. Among the latter was in the biological realm the observation which irresistibly imposes on the mind the conviction that living matter works for an end. But working for an end, and the teleology it prompts, is never a mere phenomenon. And since it is more than that, phenomenology can cope with it only by erecting its professedly non-metaphysical status into a pseudo-professional metaphysics.

It is in this light that one should see Maritain's mixed feelings

about the rise of criticism in the 1920s against the purely material-
istic biology of the late nineteenth century. Appreciative as he was
of that reaction he saw an even more dangerous irrationalism
entering the scene in its wake. In fact he saw that irrationalism
fomented by what he called the "re-entry of intellect into science."[69] He had in mind the ever more fashionable philosophizing
by scientists in terms of phenomenology. Not science, he warned,
but "only good philosophy can drive out bad philosophy."[70] Insisting as he did that the future, nay, the salvation of biology lies in a
full attention given to physico-chemical as well as to the intuitive
account, he urged that the interpretation of such intuition be
assigned to a separate field, the philosophy of nature.[71]

Such was a stance in full harmony with his basic contention that
unless one has carefully distinguished, unless specific roles have
been carefully kept distinct, their unification would result in confusion. The deluge of words that for the past three decades has been
poured out on the unification of two cultures, scientific and humanistic, fully justifies Maritain's concern. For if anything has been
produced by that deluge it is the steady dilution of humanities.
Humanities have either been turned into mysticism as was done in
Polanyi's doctrine of personal knowledge in which science too
becomes ultimately mystical, or they have been turned into mere
epiphenomena as shown by the writings of Bronowski, or they
were merely given charming literary lip-service as illustrated by
the oracular utterances of the late high-priest of two cultures, C. P.
Snow. The common delusion of all these efforts is the inability of
their proponents, or their lack of courage, to recognize and to pro-
claim that science is a limited knowledge.

Once Maritain had seen the intellectual light in which he found
full meaning, he no longer suffered from that inability or lacked
that courage. It took enormous courage as it still does today, to
say, as he did in that first article of his, that science was a
"diminished knowledge" whose accuracy grew in the measure in
which its objective becomes more and more restricted. It took and
still takes enormous courage to stand up to scientism because one
becomes thereby, to quote from Maritain's first article, as isolated
in modern culture as "were the few just men in Sodom and
Gomorrha."[72] It took and still takes courage to call upon reason to
protest "against nine-tenths of what arrives to the public in the
name of science."[73] It took and still takes courage to state that

there would be no opposition between faith and science if science were to be of philosophical "good faith."[74]

Yet it is on a point closely related to that harmony between faith and science, a harmony all too dear to Maritain, that a puzzling question arises concerning his interpretation of science in general and his survey of the new physics in particular as both stood around 1930. By then the expansion of the universe or rather the recessional velocity of galaxies was a widely discussed topic. The expansion of the universe in turn was universally connected with Einstein's general theory of relativity. Maritain himself could not be unaware of the shock which Einstein created in 1922 in the Sorbonne by insisting on the finiteness of the universe as the most acceptable among various possibilities offered by his cosmology.[75] Yet Maritain failed to mention either the finiteness of the universe or its expansion, though he spoke of the universal increase of entropy and even allowed it to be a somewhat imperfect but valid pointer to an absolute origin.[76] What is especially puzzling is that Maritain nowhere speaks to any notable extent of the universe as such.[77] It is impossible to assume that he was unaware of the true aim of Kant's disparagement of the notion of the universe, the very gist of the Kantian criticism of the proofs of the existence of God. Consequently, Maritain should have more than welcomed the twentieth-century discovery of the universe by the new physics.

He did not, and the fact is perplexing for two reasons. One is the ever stronger reliability of the confidence that science has indeed become a valid discourse about the quantitative interaction among the totality of material things, or the universe. Maritain did not make prognostications about the special directions of the future progress of science, but he expected the unveiling by science of the quantitative aspects of things on an ever larger scale. As this process unfolds, the material universe shows itself ever more specific. The consequences in this respect of the 3°K cosmic background radiation are simply breathtaking. But this overall cosmic specificity which manifests itself in all parts, big and small, early and late, of the universe, is precisely the ground on which the critical realist philosophy has always staked its basic claims.[78] Herein lies the second of those reasons concerning Maritain's puzzling silence. For it is the specificity of things which is their most gripping pointer to their existence and also to their possibility of existing in some other way, and if so, to their possible non-existence or rather non-necessary existence, and therefore radical contingency. Maritain, if I may conclude on a slight note of criticism, did not seem to realize how right he was when he spoke of the progress of new physics as one of the most beautiful and moving spectacles the modern mind is privileged to contemplate.

[1]J. Maritain, *Distinguish to Unite or The Degrees of Knowledge,* newly translated from the fourth French edition under the supervision of Gerald B. Phelan (New York, 1959), p. 154.

[2]The French original was first published in 1932.

[3]*Degrees of Knowledge,* p. 155.

[4]If metaphysics is indeed a move, nay a jump, beyond physics, its soundness should seem to depend on the measure in which the metaphysician's hold on the jumping board is firm and secure. That hold is inconceivable without a fair grasp of physics, the par excellence study of empirical nature or *physis.*

[5]For example, Jacob Bernoulli, Leonard Euler, Jacques Ozanam. For further details, see my *The Relevance of Physics* (Chicago: University of Chicago Press, 1967), p. 462.

[6]Or its logical fruit, *Naturphilosophie.* Because of the campaign led by *Naturphilosophes,* in the 1820s and 1830s, against genuine scientists, such as George Ohm, the latter did not spare the former of well-merited invectives. For further details, see ibid., p. 334.

[7]J. C. Maxwell, "Address to the Mathematical and Physical Section of the British Association" (1870), in *The Scientific Papers of James Clerk Maxwell,* ed. W. D. Niven (Cambridge, 1890), vol. II, p. 216.

[8]M. Born, *Atomic Physics* (1935; 6th ed., New York, 1957), p. 312.

[9]I refer to Professor Robert H. Dicke of Princeton University.

[10]No wonder. Those statements collected by Prof. G. Holton in pioneering studies, could but lose their impressiveness by his and also Einstein's efforts to secure them a purely "rationalist" character. For details, see my Gifford Lectures, *The Road of Science and the Ways to God* (Chicago: University of Chicago Press, 1978), pp. 185 and 193.

[11]R. Maritain, *We Have Been Friends Together: Memoirs,* tr. J. Kernan (New York, 1942).

[12]Ibid., p. 67.

[13]Ibid., p. 64.

[14]Ibid., p. 39. The list, as far as the names are concerned, is accurate.

[15]See *Bulletin administratif du ministère de l'instruction publique. Tome XXXVII Année 1885, Nos. 630-654* (Paris: Imprimerie Nationale, 1885), pp. 219-220.

[16]It is my pleasant duty to express my gratitude to Mme Rivaud, director of

the Library of Lycée Henri IV, for making that registry and other documents available to me.

[17]"Très intelligent, un peu distrait, n'est pas assez écolier; pourrait encore mieux faire en ne pensant pas à autre chose qu'à la classe."

[18]Edet (1854-1903) was also giving courses in Latin composition at the Sorbonne and is mentioned by Raissa (*We Have Been Friends Together*, p. 69) as the "superlative Latin scholar whose corrections of Latin themes were such masterpieces that the students eagerly vied to get hold of them."

[19]"Excellent élève, d'esprit parfois aventureux, mais fin et distingué, il a beaucoup gagné sous le rapport de la méthode" (1st trimester); "élève distingué, tout à fait digne d'estime et de sympathie" (3d trimester). It tells something of young Jacques' mischievousness that apparently he kept recalling later a speech defect of Dereux to Raissa, for whom Dereux was Jacques' only teacher at Henri IV to be mentioned by name: Dereux "who was called Dereuf because he always added an *f* to the ends of words" (*We Have Been Friends Together*, p. 67).

[20]*Année scolaire 1899-1900. Lycée Henri IV. Distribution solennelle des prix faite le 29 juillet 1899* (Paris: Imprimerie typographique de P. Dubreuil, 1900), p. 52.

[21]See Dossier AJ[16] 4785 (p. 489) in the Archives Nationales (Paris). There is no similar documentation available for the procès-verbaux for *licences ès sciences* for the years 1899-1919. I would like to express my appreciation to Mr. Laurent Morelle, archivist of the Sorbonne, for his kind help in guiding me to this and other documents relating to Maritain's years there.

[22]Or *fiche scolaire*. See Dossier AJ[16] 5712 in the Archives Nationales. The only information on the *fiche scolaire* of Mlle Oumancoff, Raice (AJ[16] 5716) is that she failed (éliminée) in the three examinations which she took in geology (Oct. 1901), in botany (Oct. 1901) and again in geology (July 1902).

[23]*We Have Been Friends Together*, pp. 64-65.

[24]Among the publications of Le Dantec (1869-1917) are such titles as *L'athéisme* (1906), *La lutte universelle* (1906), *Contre la métaphysique* (1912) in witness to the antimetaphysical creed into which Le Dantec turned Darwinism, in itself a mere scientific method.

[25]See *We Have Been Friends Together*, pp. 72-78.

[26]Hans A. E. Driesch (1867-1941) was at first a convinced mechanist, a fact which made his turning to vitalism, around 1894, all the more significant and newsworthy.

[27]"Le néo-vitalisme en Allemagne et le Darwinisme," *Revue de philosophie*, 17 (1910), 417-441.

[28]H. Driesch, *La philosophie de l'organisme*, traduit de l'allemand par M. Kollmann, préface de J. Maritain (Paris, 1921), pp. i-xi.

[29]H. Driesch, *The Science and Philosophy of the Organism* (London, 1907-08). Driesch wrote and delivered these lectures in English of which he had a full command.

[30]Jacques and Raissa returned "for good" from Heidelberg in May 1908. They first went there on August 25th 1906 (see *We Have Been Friends Together*, pp. 180 and 198). Their stay in Heidelberg was made possible by a grant from the Michonis Fund.

[31]A. France, *The Opinions of Jerome Coignard*, tr. W. Jackson, in *The Works of Anatole France*, vol. IV (New York, 1924), p. 114.

[32]France made that remark in the late 1880s in his review of Flammarion's once famous popularization of astronomy, *Uranie*. See A. France, *La vie littéraire* (Paris, 1888-92), vol. III, p. 212.

[33]Brunetière's phrase could but incense the scientistic establishment of the Third Republic for more than one reason. First, Brunetière had made his scholarly reputation with a study in which he interpreted the development of *belles-lettres* in terms of Comte's law of three phases, a widely accepted basis of scientism. Second, Brunetière's position as editor-in-chief of the *Revue des Deux Mondes* assured him a world-wide hearing. Third, Brunetière's stated preferences for talking of "the bankruptcy of science" instead of seeing in it the hope for the future appeared in an article in which he described the Vatican as that very hope, following an audience with Leo XIII ("Après une visite au Vatican," *Revue des Deux Mondes*, 127 [1895], 97-118; see p. 98).

[34]The *Dictionnaire alphabétique et analogique de la langue française* Part 7 R-Z (Paris:Presses Universitaires de France, 1951-1970, p. 358), a work as authoritative as the big *Oxford Dictionary of the English Language*, is in conflict with the truth of Raissa's remark that Jacques coined the word scientism (*We Have Been Friends Together*, p. 77). In the *Dictionnaire*, 1911 is given as the first appearance of "scientism" and Le Dantec, of all people, is given the credit. In view of the friendship between the Maritains and Le Dantec, the latter was most likely the recipient of a reprint of Jacques' article published a year earlier. One cannot help thinking of Chesterton's dictum about Catholics: "to be first and to be forgotten."

[35]J. Maritain, "La science moderne et la raison," *Revue de philosophie*, 16 (1910), 575-603. That Maritain attributed some importance to that article is evident from its having been selected by him as the first chapter in his *Antimoderne* (1922) which appeared in German translation in 1930. This is, of course, not the only detail which has readily come to my attention from *The Achievement of Jacques and Raissa Maritain: A Bibliography 1906-1961* (Garden City, N.Y.: Doubleday, 1962), an epitome of meticulous scholarship, for which any student of Maritain should feel much indebted to its authors, Donald and Idella Gallagher.

[36]*We Have Been Friends Together*, pp. 63-64.

[37]Ibid., pp. 39-40.

[38]Since the publication in 1970 of Hegel's *Enzyklopädie* in a three-volume English translation by M. J. Petry (*Hegel's Philosophy of Nature* [London]),

none of the countless Hegel-experts who read only English can pretend the unavailability of a work in which Hegel provided a far more destructive exposure of his own philosophy than any critic of it could ever write. For some eye-opening details, see my *Angels, Apes, and Men* (La Salle, Ill.: Sherwood Sugden and Company, 1983), pp. 37-38.

[39]As argued in the respective chapters of my Gifford Lectures (see note 10 above).

[40]*Degrees of Knowledge*, p. 21.

[41]Maritain did not seem to know about the series of articles which Duhem published between 1891-94 in the *Revue des questions scientifiques*. There Duhem put a much greater emphasis than he did in his classic work, *La théorie physique* (1906), on a realist epistemology (metaphysics) which alone can make meaningful the positivist methodology of science. It is in this corrective light that one should read Maritain's strictures (*Degrees of Knowledge*, p. 44) of Duhem's theory of scientific knowledge. For further details, see my book, *Uneasy Genius: The Life and Work of Pierre Duhem* (Dordrecht: Nijhoff, 1984), pp. 321-22.

[42]The following list is gathered from the footnotes of the *Degrees of Knowledge*.

[43]F. Gonseth, *Les fondements des mathématiques* (Paris, 1926).

[44]See *Degrees of Knowledge*, p. 183.

[45]Ibid., p. 141.

[46]Ibid., pp. 140-41.

[47]Ibid., p. 140.

[48]Ibid., p. 156. Maritain was not disturbed at all by the fact that he had found those statements quoted by Langevin, a leading French physicist and also a professed Marxist. Precisely because Marxism insisted on the reality of matter as a primary datum, Maritain (as was the case with Gilson as well) saw in it a lesser danger than in the phenomenologists' methodical avoidance of questions relating to reality as such.

[49]*Degrees of Knowledge*, p. 158. There is a complete procès-verbaux of the question-answer period that followed Einstein's lecture on April 6, 1922, before the *Société française de philosophie* in its *Bulletin*, Tome 17 (1922), 91-113.

[50]This perspective of Einstein's physics, mentioned but fleetingly by Maritain (*Degrees of Knowledge*, p. 157), is developed in ch. xii of my Gifford Lectures (see note 10 above) and in my paper, "The Absolute Beneath the Relative: Reflections on Einstein's Theories," *Intercollegiate Review* 20 (Spring/Summer, 1985), pp. 29-38.

[51]The only major exception was Planck whose lectures, widely read in Germany, did not apparently come to Maritain's notice.

[52]For details and documentation, see "Chance or Reality: Interaction in Nature versus Measurement in Physics," reprinted here as Ch. 1.

[53]*Degrees of Knowledge,* p. 160.

[54]Ibid., p. 189.

[55]Details are given in my "Chance or Reality."

[56]See *Degrees of Knowledge,* pp. 165-171.

[57]Ibid., pp. 151-52 and 181-84.

[58]Most appropriately some of those strictures (*Degrees of Knowledge,* pp. 183 and 188) were administered in a section which Maritain entitled: "Dangerous Liaisons."

[59]For a discussion, see ch. xv, "Paradigms or Paradigm," in my Gifford Lectures.

[60]*Degrees of Knowledge,* p. 189.

[61]Ibid., p. 190.

[62]Ibid., p. 191.

[63]He did so in his Presidential Address to the Royal Society in 1981. See *Supplement to Royal Society News,* Issue 12, November (1983), p. v.

[64]*Degrees of Knowledge,* pp. 195-96 and 198.

[65]Ibid., pp. 192-93.

[66]Ibid., p. 198.

[67]Ibid., p. 192.

[68]Ibid., p. 197.

[69]Ibid., p. 197.

[70]Ibid., p. 191.

[71]This is also the gist of the octogenarian Gilson's book, *D'Aristôte à Darwin et retour* (1974) published in J. Lyon's English translation, *From Aristotle to Darwin and Back Again,* by the University of Notre Dame Press, 1984.

[72]"La Science moderne et la raison," p. 599.

[73]Ibid., p. 587.

[74]Ibid., p. 603.

[75]For details on that shock, see my *The Paradox of Olbers' Paradox* (New York: Herder and Herder, 1969), pp. 224-25.

[76]*Degrees of Knowledge,* p. 187.

[77]In *The Degrees of Knowledge* the few references to the universe (pp. 130, 132, 137, 187) are fleeting indeed.

[78]The contributions which modern scientific cosmology can make to philosophical cosmology and natural theology are set forth in my article, "From Scientific Cosmology to a Created Universe," *Irish Astronomical Journal* 15 (March 1982), 253-262.

4

Chesterton's Landmark Year: The Blatchford-Chesterton Debate of 1903-1904

For all its vastness and variety, the Chesterton literature contains no study on the debate which Chesterton had with Blatchford in 1903-1904, although it was recalled in Chesterton's *Autobiography* as a "landmark in my life."[1] No other event received a higher encomium from Chesterton the autobiographer. Among the many opponents and critics Chesterton encountered, Blatchford was esteemed just as highly. Or as Chesterton wrote in the early 1930s to Laurence Thompson, already at work on a biography of Blatchford to be published in 1951:

> Very few intellectual swords have left such a mark on our time, have
> cut so deep, or remained so clean. His [Blatchford's] case for

This article is based on a paper delivered at the Conference, "Gilbert Keith Chesterton 1874-1936: An Interdisciplinary Approach," at the Catholic University of America, Washington, D.C., March 10, 1984. Reprinted with permission from *The Chesterton Review* 10 (1984), pp. 409-23.

Socialism, so far as it goes, is so clear and simple that anyone could understand it, when it was properly put. In other words, Blatchford is an artist if ever there was one in this world. His triumphs were triumphs of strong style, native pathos and picturesque metaphor; his very lucidity a generous sympathy with simpler minds. For the rest, he has triumphed by being honest and by not being afraid.[2]

Bent as he was on exaggerations, which are both his strength and weakness, Chesterton must have had reasons for putting on so high a pedestal a writer now completely forgotten. Chesterton's chief interest in Blatchford related, of course, not to his style but to the message it carried. Chesterton's reaction to that message should have a paramount interest for Chestertonians because of the catalytic role it played in the emergence of *the* Chesterton for whom he was taken for the rest of his life.

Following that debate Chesterton was no longer looked upon as a mere wizard of paradoxes whom one can read with nonchalant enjoyment. Through that debate, he emerged as an uneasy wizard who had to feed on paradoxes because of his total commitment to a most paradoxical philosophy, or plain Christian creed. Blatchford himself could never bring himself to the point of believing that a man with Chesterton's mental powers could have really believed in that creed. To the very end of a very long life, Blatchford (who died in 1943 at the venerable age of ninety-two) held that Chesterton, though not a poseur or imposter, was a consummate actor who played so well the role of believing Christian dogmas as to become convinced in the end that he really believed in them. "Like you," Blatchford wrote to a friend long after his debate with Chesterton, "I don't understand Chesterton turning a Catholic. But he was a subtle thinker, and I have noticed that your subtle thinker often loses his way in his own intricacy. He spins a kind of web round his own brain."[3] At any rate, it was in the heat of his debate with Blatchford that Chesterton first became aware of being suspected of insincerity. That suspicion followed him from various quarters to the end of his life and continued well after his death.

The Blatchford-Chesterton debate prompted Chesterton to not a few utterances that have a perennial freshness without being loaded with the kind of metaphysics that is bound to produce uneasiness, though never in the incurably facile mind. The appetite of even the non-believing Chestertonians should have been whetted for knowing more on the subject on reading in *Heretics* that hilariously serious account of Blatchford's editorial skill:

> The whole modern world is pining for a genuinely sensational journalism. This has been discovered by that very able and honest journalist Mr. Blatchford, who started his campaign against Christianity, warned on all sides, I believe, that it would ruin his paper, but who continued from an honourable sense of intellectual responsibility. He discovered, however, that while he had undoubtedly shocked his readers, he had also greatly advanced his newspaper. It was bought —first, by all the people, who agreed with him and wanted to read it; and secondly, by all the people who disagreed with him, and wanted to write him letters. Those letters were voluminous (I helped, I am glad to say, to swell their volume), and they were generally inserted with a generous fulness. Thus was accidentally discovered (like the steam engine) the great journalistic maxim—that if an editor can only make people angry enough, they will write half his newspaper for him for nothing.[4]

Honourable and *generous* are the key words in the foregoing quotation. They echoed Chesterton's characterisation of Blatchford's decision as "a gesture of unparalleled generosity" whereby the columns of *Clarion* were offered to critics of his ideology.[5]

Generosity and honourableness were certainly evident in Blatchford from his youth, but nothing indicated that young Robert Blatchford, apprenticed to a brushmaker at the age of fourteen, would turn into by far the most widely-read author in England at a time, the turn of the century, which witnessed the meteoric rise of a Wilde, a Shaw, and a Wells. The son of provincial actors, Blatchford lost his father at the age of two. Out of school by the time he turned a teenager, young Blatchford met, as a brushmaker's apprentice, Sarah Crossley whom he was to marry in 1880. To win her hand he had to raise his status, a long and arduous project, which brought out the very best in Blatchford's character. For seven years (1867-1873), Blatchford served in the 103rd Regiment in India, spending all his free time, to the amazement of his fellow recruits, in improving his English. He did so by reading voraciously and by writing countless long letters to Sarah. He also read so much popular science that, upon his return to England, he succeeded in getting a job as a time-keeper with a navigational company in Norwich. But he sensed that his future lay with writing. His first story was sold in 1883 and, shortly afterwards, he was on the staff of *Sunday Chronicle* (Manchester) with an annual salary of a thousand pounds! In sensing his literary powers and in the grip of a higher mission, he wanted his own forum, which he obtained by purchasing the struggling Socialist weekly, *Clarion*. Its circulation soon rose, on the strength of Blatchford's essays published there, to the impressive figure of 60,000 copies. Even more spectacularly successful was a collection of Blatchford's essays from *Clarion* published in 1893 under the title *Merrie England,* a gospel of socialist Utopianism.

That *Merrie England* sold in two million copies in a mere decade was due only in part to Blatchford's simple and lucid style. The socialist Utopia into which Blatchford wanted to turn an unhappy England was based on that commodity, science, which had just proved itself the most effective selling device of modern times. Here, too, Blatchford's lucid style and simple diction were of enormous help. The long Victorian sentences, in which Herbert Spencer had, during the preceding decades, sung the praises of science as the magic key to all values, ranging from sanitation to sanity, from factories to families, from playthings to politics, from medicine to morals,[6] were broken down by Blatchford into phrases hardly ever longer than ten simple Anglo-Saxon words. His paragraphs were never longer than five lines. Nor were Blatchford's readers supposed to know even that much of popularised science which T. H. Huxley's famed lay-sermons contained. Not that Blatchford had not constantly recommended those standard popularisations of science to his readers. Undoubtedly most of them felt convinced that, having understood Blatchford's extreme simplification of science, they had mastered all the sciences. Few, if any of them, had realised that what they had assimilated from Blatchford's writings was not science but scientism. The word *scientism* still had to enter the English language—and, as one might have expected, from French books, following its first use by Jacques Maritain in 1910.[7] Scientism was to denote that utterly superficial belief, already widely adhered to in the England of Spencer and Huxley, that the empirico-quantitative method was the only reliable access to reality and truth. Anything else grasped by any other method— intuitive, aesthetic, philosophical, let alone metaphysical and religious—was either a mere epiphenomenon or a mere illusion. One chief victim of scientism was free will, whose vindication became a principal concern for Chesterton, partly through his debate with Blatchford, a champion of scientistic determinism. In trying to see through the fallacies of scientistic determinism, Chesterton had to reflect on the nature of determinism in science. The result was a wealth of insightful dicta on scientific law, on evolution, on universe, which show Chesterton as an outstanding interpreter of science.[8]

It should not, therefore, be surprising that shortly before Chesterton referred in the *Autobiography* to his debate with Blatchford as a landmark in his life, he spoke of Darwinism as one of the chief carriers of the modern intellectual disease which, in the name of science, tries to undermine man's belief in his own freedom and responsibility. In the same context, there is a pointed remark on the books of the German biologist, Ernst Haeckel, enormously popular in England in the closing decades of the nineteenth century. Haeckel's two-volume *History of Creation,* put in English hands in six editions between 1876 and 1914, was a chief source for Blatchford and for many of his compatriots, who tried to appear knowledgeable about science while pushing the cause of scientism, a thing very different from science. Science, or rather scientism, is all too obvious in the four places where Blatchford is mentioned in *Orthodoxy.*[9] Blatchford's appearance in *What's Wrong with the World* is in connection with eugenics,[10] a patently scientific, or rather, scientistic topic.

But to return to Blatchford whose next collection of essays, *God and My Neighbour,* a dogmatics of socialism, published in 1903, was only slightly less successful than *Merrie England.*[11] It should not be remembered today, except for the appearance on the scene at the same time of "the comparatively young though relatively rising journalist." Such was Chesterton's self-portrait relative to that year. Indeed, he was rising far above most of his contemporaries. Only Chesterton's name would ring a bell today with the average learned man as he read through the table of contents of *The Religious Doubts of Democracy,*[12] a collection of two dozen essays from *Clarion,* each about 1500 words long, by a dozen or so authors, who were selected by George Haw, a chief aid of Blatchford. The writer most often invited to vent in *Clarion* his disagreement with Blatchford was Chesterton, "the comparatively young though relatively rising journalist."

The reason for this was in part Chesterton's impressive productivity. *Greybeards at Play* (1900) and *The Wild Knight* (1900) made

Chesterton known as a poet. As an essayist, he made a name for himself with *The Defendant* (1901) and *Twelve Types* (1902). In May 1903 there followed *Robert Browning* which earned him a reputation as a literary critic. By then Blatchford was prompted by a remark of Chesterton in the March 14 issue of *The Daily News* to a brief comment about him in *Clarion* (March 27). Still another remark of Chesterton (*Daily News,* April 4) made Blatchford decide on a broadside "Wolf! Wolf!! Wolf!!!" (*Clarion,* July 31) in which Chesterton figured as the chief of "Free-Willers." With that, the battle between the two raged in full. Chesterton's "Mr. Blatchford and Free Will" (*Clarion,* August 7) was followed by Blatchford's "Love and Hate" (August 14), which in turn made Chesterton reply with another "Mr. Blatchford and Free Will" (August 27) where he exposed Blatchford as a champion of that eastern fatalism which Christendom had pushed back well beyond the banks of the Danube. Chesterton left unanswered Blatchford's immediate rejoinder, "The Battle of the Danube" (September 4).[13]

Not that Blatchford had disappeared from Chesterton's horizon. In early November, he reproached Blatchford in the Christian socialist monthly, *Commonwealth,* for his "aggressive infidelity" on the ground that it obviously contradicted Blatchford's championing determinism. Determinism, Chesterton argued the obvious, left no room for responsibility for any crime and no room, for that matter, for exhortation (crusading or not) against crime and on behalf of virtue. By the time Blatchford replied with his "Clerical Logic" (*Clarion,* November 13), Chesterton's review of *God and My Neighbour* was being typeset for the next-day issue of *The Daily News.*

No one modestly familiar with Chesterton needs to be convinced that, like anything Chesterton wrote, his criticisms of Blatchford are full of sparkling phrases, time and again loaded with philosophical profundity, tied to apparently most trivial human actions and dicta. Thus he reminded Blatchford that on the basis of his determinism, staple phrases, such as "Please pass the mustard," are contradictory because "please" implied free will on the part of the one who was asked to comply. Right there and then, Chesterton connected human sanity, pre-empted by Blatchford's determinism, with Christianity as its sole safeguard:

She [Christianity] knows that as soon as you want any conceivable human reality, if it be only to say "Thank you" for the mustard, you will be forced to return to her hypotheses, where she sits, guarding through the ages the secret of an eternal sanity.

And when Blatchford retorted that Christianity is to be blamed for determinism because she makes God create any human action, Chesterton reminded Blatchford of two things: One was that "physical discussions can be definitely ended; philosophical ones can only be reduced to first principles." Freedom of the will was one such principle on which rested that very consistency which created human civilisation. Thus, and this was Chesterton's second point, if Christianity insisted on free will, it was only because she considered it to be God's free gift to man, the kind of gift which does not lose its consistency by passing from the giver to the recipient. And so it was with any blame and praise, a point which Chesterton drove home with an eye on the presumed policy in the editorial offices of *Clarion*. Chesterton could not believe that any clerk had ever been scolded there by the editor with the following words: "My blameless Ruggles, the anger of God against you has once more driven you, a helpless victim, to put your boots on my desk and upset the ink on the ledger. Let us weep together." For, if indeed such was the procedure there, then, Chesterton added, "gaily will one I know apply for the next vacancy in that philosophical establishment."

Yet the very fact that Chesterton, in these unsolicited letters to the owner and editor of *Clarion*, had to tack his remarks to particular statements of his opponent, forced him to argue *ad hominem*. While this made possible a gripping immediateness, it was not the format for developing a theme. Chesterton was shifting toward the thematic when he reviewed Blatchford's *God and My Neighbor*. The thrust of his dissecting that book was that determinism was simplistic whereas human reality was immensely complex. Complexity in turn reflected the presence of deep paradoxes which served as the unfailing source for ever-fresh rise of man's religious orientation:

What is primary? The primary thing is plain human experience. It is the thing out of which religion necessarily arose. . . . Religion arose because there are incurable contradictions, impossible paradoxes in existence itself.

Chesterton meant to deal in another article with other details in Blatchford's book that cried out for a recognition of the "practical paradox of the nature of Faith, that great paradox, lasting through the ages, which makes faith always nonsensical and always necessary." That second review was never written, most likely because around the end of 1903 Blatchford decided to open *Clarion* to writers, to be chosen by Haw, who would present systematic rebuttals of Blatchford's philosophy. Chesterton's turn came in July and August in four articles which were so many rehearsals for the thematic chapters of *Heretics* and especially of *Orthodoxy*. Had Chesterton tried to sound in those four essays as a professional philosopher, or to do so in his subsequent writings, he would today be just as forgotten as Blatchford is. One has to be oneself in order to sound true and to make impact. The secret of the impact which "the relatively rising journalist" made—and was to make for the rest of his life—was nothing else than the fact that just around the turn of the century Chesterton had found himself. It was an agonising find, preceded by the fear of losing oneself in sheer idealism and solipsism, or of losing all hold on reality. The story has been told many times, above all by Chesterton himself, to be repeated here however briefly.

It was Chesterton's newly won hold on reality which imposed on him the only strategy of effectively countering Blatchford's argument. The strategy consisted in unfolding in full Blatchford's chosen perspective and making his principal arguments, or Blatchford's four big guns, appear in its light. Once this was done, Chesterton could argue: "Gentlemen of the Secularist Guard, fire first!"[14] Those four guns were in turn the pagan myths paralleling Christian myths, the alleged gloominess of Christianity, the breeding by Christianity of cruelty and tumult, and the local or particular origins of Christianity (and of Judaism). As to the first, Chesterton noted that it discredited anything that was seen and experienced by many, an outcome certainly paradoxical in the sense of being self-defeating. Yet, as Chesterton added with contempt, it was on such basis that his opponent tried to "convict the wretched G.K.C. of paradox." The charge of gloominess was turned around by Chesterton with a reference to the saints giving up ordinary joys for the sake of a most joyful mystical experience, systematically ignored by secularists. "Whose is the paradox?" Chesterton exclaimed "submissively." As to the cruelties fomented by Christianity, Chesterton put them in the perspective of cruelties and tumults triggered by secularism:

> The mere flinging of the polished pebble of Republican Idealism into an artificial lake of eighteenth-century Europe produced a splash that seemed to splash the heavens and a storm that drowned ten thousand men. . . . Men swept a city with the guillotine, a continent with the sabre, because Liberty, Equality, and Fraternity were too precious to be lost.

Clearly, it was right to counter the objection that Christianity produced tumults with the single word: "Naturally." As to the particular origins of Christianity and Judaism, Blatchford's secularist scientism called for a Moses encountering Energy instead of the burning bush. Would Moses thereby have become more genuine and credible, let alone to Secularists?

In vindicating the right perspective, Chesterton had to find more than common sense: Reality, he felt, could be safeguarded only by finding Christianity, nay, the very divinity of Christ. In that respect too, the essays written against Blatchford represent a landmark in Chesterton's mental and spiritual development. Chesterton's belief in the divinity of Christ is stated there five years before *Orthodoxy* and twenty years before *The Everlasting Man*, a book which, let us not forget, is above all about the everlasting God made Man. Indeed Chesterton's second Blatchford essay carried the title "Why I believe in Christianity,"[15] whereas the third, "Miracles and Modern Civilisation,"[16] ended with a powerful emphasis on the divinity of Christ, the very miracle which makes all miracles possible and meaningful:

> The Romans were quite willing to admit that Christ was a God. What they denied was that He was the God—the highest truth of the cosmos. And this is the only point worth discussing about Christianity.

The fourth of the Blatchford essays, "The Eternal Heroism of the Slums,"[17] also fully anticipates Chesterton the popularist who knew poverty at close range. His eyes were never shut to the slums only a few blocks from his house in Battersea. Nor was he unfamiliar with the slum dwellers. This is why he thundered at Blatchford, according to whom good morals were a function of comfortable living, for which Blatchford's *Merrie England* was supposed to provide the blueprint. Blatchford's association, Chesterton cried out,

> of vice and poverty is the vilest and the oldest and the dirtiest of all the stones that insolence has ever flung at the poor. . . . I will not deign to answer even Mr. Blatchford when he asks "how" a man born in filth and sin can live a noble life. I know so many who are doing it, within a stone's throw of my own house in Battersea, that I care little how it is done. Man has something in him always which is not conquered by conditions. Yes, there is a liberty that has never been chained.

The question of liberty versus determinism provides the logic that runs across the arguments that had led Chesterton in his second and third essays to his declaration of the divinity of Christ. There, too, he shows himself to be the kind of thinker whose emergence has so far been postdated in all Chesterton studies. Take, for instance, that famed chapter in *Orthodoxy,* "The Ethics of Elfland," which (and this, too, Chestertonians have invariably ignored) became, forty years after its publication, a classic in the modern philosophy of science.[18] All that penetrating analysis by which Chesterton laid bare the utter dependence on metaphysics of that much-vaunted notion of scientific law, or laws of nature, had already been set forth in inimitable phrases in the third Blatchford essay. Introductory to that essay was his merciless exposure of the philosophy of determinism. Needless to say, any self-styled determinist is deliberately a determinist and, therefore, an easy target for well-deserved ridicule. Chesterton did something more. He showed that the real threat of determinist materialism is not in its self-defeating logic which no determinist has ever been known to obey even to a modest degree. The real danger lay in that pleasant-looking half-way edifice erected on a road leading to an absolute dead-end or "a complete darkness," to quote Chesterton. Had he lived in our pollution-conscious society, he would have opposed determinism—sociological, physical, economic, genetic, behaviorist, Marxist, Freudian, or what not—with a crusade in defense of clean air—intellectual and spiritual. Instead, he spoke of the darkness brought about by a determinism claiming to itself the bright light of science: "The determinist makes the matter of the will logical and lucid; and in the light of that lucidity all things are darkened, words have no meaning, actions no aim." And certainly not the aim and meaning of science. Why is it, Chesterton asked, that the Christian defense of the mystery of the freedom of the will went hand in hand with scientific discoveries? "The East has logic and lives on rice. Christendom has mysteries—and motor cars." He put a vast truth in a nutshell. Similar was his question "Are you surprised that the same civilisation which believed in the Trinity, discovered the steam engine?"[19]

The third essay, "Miracles and Modern Civilisation," was largely devoted to the theme that instead of assuring the repetition of an event, scientific laws presuppose it as a natural miracle. Furthermore, that natural miracle was understandable only if there was a Divine Will. From the Divine Will, Chesterton, always in the grip of the concrete, proceeded to its most concrete manifestation, Christ (in phrases already quoted). Of course, that Will was not a capricious will, but a will it was and still is, and without it the whole status of natural law or scientific law was a mere begging of the question. To say this openly was already cultural lèse majesté and even more so when said with the devastating conciseness of which few could ever deliver themselves so promptly and so often as Chesterton:

> The question of miracles is merely this: Do you know why a pumpkin goes on being a pumpkin? If you do not, you cannot possibly tell whether a pumpkin could turn into a coach or couldn't. That is all. All the other scientific expressions you are in the habit of using at breakfast are words and winds.[20]

What was the most likely reaction to this fearless demolition of science as the chief ideal of secularism, and of the no less courageous erecting in its place the statue of Christ? Chesterton introduced his autobiographical account of the Blatchford debate with a graphic portrayal of that reaction. It made unforgettable for him a dinner thrown by the staff of *Clarion* to those invited to debate in its columns its editor's philosophy:

> But I remember that there was, sitting next to me at this dinner, one of those very refined and rather academic gentlemen from Cambridge who seemed to form so considerable a section of the rugged stalwarts of Labour. There was a cloud on his brow, as if he were beginning to be puzzled about something; and he said suddenly, with abrupt civility, "Excuse my asking, Mr. Chesterton, of course I shall

understand if you prefer not to answer, and I shan't think any the
worse of it, you know, even if it's true. But I suppose I'm right in
thinking you don't really *believe* in those things you're defending
against Blatchford?" I informed him with adamantine gravity that I
did most definitely believe in those things I was defending against
Blatchford. His cold and refined face did not move a visible muscle;
and yet I knew in some fashion it had completely altered. "Oh, you
do," he said, "I beg your pardon. Thank you. That's all I wanted to
know." And he went on eating his (probably vegetarian) meal. But I
was sure that for the rest of the evening, despite his calm, he felt as if
he were sitting next to a fabulous griffin.[21]

That condescending "Oh, you do. I beg your pardon. Thank you.
That is all I wanted to know" has become in our times the gravest
concern for many Catholic intellectuals (and even of some Chestertonians) craving the applause of the establishment.

Many are most eager to play on paradoxes provided they are not
taken for more than witty paradoxes. Chesterton was taken to task
for his paradoxes largely because he showed that Blatchford, bent
on convincing "that wretched G.K.C. of paradoxes," was guilty of
a baffling oversight of his own paradoxes, so many destructive
threats to human nature itself. Chesterton took up the gauntlet,
which challenged, beyond Blatchford, the entire secularist establishment, because he had the courage to stand up for his dearest
convictions. He refused to be taken for a mere paradoxer, however
enjoyable, when he began to realise, as he put it in his first Blatchford essay, "that there are a good many people to whom my way of
speaking about these things appears like an indication that I am
flippant or imperfectly sincere."[22] The price he had to pay for his
resolve to make each and all perceive what he truly stood for has
already been hinted at by that dinner story. Here it remains to
recall Chesterton's last communication to Blatchford, duly printed
in the December 23, 1904, issue of *Clarion*. There Chesterton dis-
sected a formula offered by Blatchford as to how a little boy should
treat a capricious little girl in terms of benevolent determinism:

You say, referring to some remarks of mine about the maddening
sense of being merely a machine, that you do not feel at all like that. I
know that quite well; and I know what saves you from it. It is your
inconsistency: such inconsistency as you are prepared to exhibit in
the case of the little boy. As long as you continue, in the face of every

actual problem, to act in flat contradiction to all your principles, you will continue to be the very jolly person you are. But the world might contain some day a logical Determinist: and he would be a lunatic.

The vanishing small measure to which a determinist could act consistently with his principles was amply illustrated by Blatchford. In the wake of the storm created by his *God and My Neighbour,* he stood his ground on the basis (hardly deterministic) that "a Determinist must live up to himself."[23] No less tellingly, but much more poignantly, Blatchford became a spiritist in 1921. He ascribed this startling turnabout of his to the breaking up of the atom which, he claimed, had put an end to mechanism. He should have rather mentioned that in that very same year he lost his dearly beloved wife. A most sensitive man Blatchford certainly was, but so was Chesterton who, in addition, had the proper insights about sensitivity and about a vast array of other problems and paradoxes as well, to say nothing of things that were obvious.

[1]G. K. Chesterton, *Autobiography* (London, 1936), p. 178.

[2]L. Thompson, *Robert Blatchford: Portrait of an Englishman* (London, 1951), p. 172.

[3]L. Thompson, *Robert Blatchford,* pp. 170-171. "Chesterton," Blatchford continued, "said a very generous and handsome thing about me. And we were old opponents. But—I always felt—Chesterton was an actor."

[4]G. K. Chesterton, *Heretics* (London, 1905), pp. 117-118.

[5]L. Thompson, *Robert Blatchford,* p. 168. While this became public only years after Blatchford's death, Chesterton began his first Blatchford essay, to be discussed later, by characterising Blatchford's generosity as a "magnanimity which, like all magnanimity, is profoundly philosophical and wise."

[6]Practically all of Spencer's writings are an elaboration of his portrayal of science as a cure-all in his "What Knowledge Is Most Worth?" (1850) in *Education: Intellectual, Moral and Physical* (New York, 1889); see especially pp. 93-94.

[7]For details, see my article, "Maritain and Science," reprinted here as Ch. 3.

[8]For details and documentation see my *Chesterton, a Seer of Science* (University of Illinois Press, 1986).

[9]G. K. Chesterton, *Orthodoxy* (London, 1909), pp. 51, 215, 238-239, 256.

[10]"Popular science, like that of Mr. Blatchford, is in this matter as mild as old wives' tales. Mr. Blatchford, with colossal simplicity, explained to millions of clerks and workingmen that the mother is like a bottle of blue

beads and the father like a bottle of yellow beads; and so the child is like a bottle of mixed blue beads and yellow" is the opening shot of the chapter entitled "The Tribal Terror" in *What's Wrong with the World* (New York, 1910), p. 234.

[11] In spite of the enormous sales of these books, they can today be found only in libraries in which, because of availability of shelf-space, they were not discarded in spite of their not having been read by anyone for the past seventy-five years.

[12] In the more than 800 volumes of the *Pre-1956 Catalogue of Printed Books* held in American and Canadian libraries, only two libraries are listed with a copy of the title, *The Religious Doubts of Democracy*. The publisher was Macmillan and Company, London and New York, 1904. The name of one of the contributors, F. R. Tennant, would of course be familiar to the few specialists in natural theology.

[13] It is my pleasant duty to express my gratitude to Dr. P. E. Hodgson of Corpus Christi College (Oxford) for providing me with photocopies of the respective publications by Blatchford and Chesterton in *Clarion*. The material obtained from him is a proof of the not entirely complete listing of Chesterton's publications in the otherwise indispensable Chesterton bibliography by J. Sullivan.

[14] G. K. Chesterton in *The Religious Doubts of Democracy*, p. 21. Quotations from those four essays will be referred to their printing in this book. For the first essay, "Christianity and Rationalism," see pp. 17-21.

[15] Reprinted in *The Religious Doubts of Democracy* (pp. 61-64). In *Clarion*, the essay appeared under the title: "We Are All Agnostics until. . . ."

[16] G. K. Chesterton in *The Religious Doubts of Democracy*, pp. 87-89. The original title in *Clarion* was "Mr. Blatchford's Religion: Is Nobody To Be Responsible for Anything?"

[17] I had access to the text of that essay only in its reprinting in *The Religious Doubts of Democracy*, pp. 106-109.

[18] As shown by its being partially reprinted in *Great Essays in Science*, edited by Martin Gardner, Editor of *Scientific American* (New York, 1957), pp. 78-83, a volume in which Chesterton was on the same pedestal with Darwin, Eddington, Fermi, Einstein, and others. For further details, see my *Chesterton, a Seer of Science*, quoted above.

[19] Chesterton's early insistence on the Christian origins of modern science and technology makes him a pioneer in that respect as well. He certainly could not have learned it from the books of W. Whewell, the chief authority in England as well as elsewhere on the history and philosophy of science during the second half of the nineteenth century. The English-speaking world did not really learn about Pierre Duhem's epoch-making discoveries about mediæval anticipations of Galilean and Newtonian science until the 1920s when Christopher Dawson made several pointed references to

Duhem's learned tomes. For details, see my *Uneasy Genius: The Life and Work of Pierre Duhem* (Dordrecht, 1984), p. 414.

[20]The phrase is a clear anticipation of more than one passage in "The Ethics of Elfland."

[21]G. K. Chesterton, *Autobiography,* pp. 178-179. Almost immediately preceding this passage, Chesterton wrote: "Very nearly everybody, in the ordinary literary and journalistic world began by taking it for granted that my faith in the Christian creed was a pose or a paradox. . . . The more generous and loyal warmly maintained that it was only a joke. . . . Critics were almost entirely complimentary to what they were pleased to call my brilliant paradoxes; *until* they discovered that I really meant what I said."

[22]G. K. Chesterton, in *The Religious Doubts of Democracy,* p. 17.

[23]L. Thompson, *Robert Blatchford,* p. 173.

5

Goethe and the Physicists

An undoubtedly true sign of greatness is the persistent challenge that can be exercised by a person's thought or actions from the distance of generations, let alone centuries. Goethe is one of these persons. Not only men of letters but also scientists come across his towering figure on their intellectual journey. In an age of scientific specialization and of grave cultural splits, he looms large in the eyes of many as a universal man embodying balance and versatility. His activities were certainly most varied and relate to such different fields as literature, political administration, and last, but not least, science. It should not, therefore, be a surprise that during the last two decades or so, several leading German physicists suggested that a closer study of Goethe's ideas on science and on the science of colors in particular may well benefit our ailing, highly technological civilization.

The devastation of World War II was already engulfing much of the world when W. Heisenberg delivered a lecture on "The Teachings of Goethe and Newton on Colour in the Light of Modern

Reprinted with permission from *American Journal of Physics* 37 (1969), pp. 195-203.

Physics."[1] In it Heisenberg emphasized the cultural import of what in his view constituted Goethe's primary aim in investigations of colors, the *Farbenlehre*: the defense of the world of sense perception from the onslaught of the mathematical abstractness of physical research. As sense perception constitutes the eminently subjective realm of qualities, Goethe's primary aim can also be defined as the vindication of the subjective against the so-called objective realm of facts and procedures. In fact the outstanding Swiss physicist, W. Heitler, saw the deepest roots of Goethe's lifelong struggle against Newtonian optics in the correct realization by the poet that the line of separation between the objective and the subjective is often arbitrary and hard to define. This comment of Heitler on Goethe is in the second chapter of his perceptive study, *Man and Science,* devoted to a most urgent need: the forestalling of the further dehumanization of man by modern technology and by its spurious quantitative philosophy.[2] In Goethe's investigations on colors, Heitler saw convincing evidence that "it was quite possible to take the view that color belongs to the objective, external world."[3]

This last remark of Heitler was in substance echoed by Max Born, in his discussion[4] of Eckart Heimendahl's notable attempt[5] to give a systematic account of man's color perception from the physical as well as the psychological viewpoint including even the emotional symbolism of colors. Born, in fact, called on his fellow physicists to transcend the limitations of the quantitative method and to cultivate a receptiveness for those aspects of reality that are inaccessible to science. With specialization on the rampage, Born warned, the whole scientific enterprise has become a senseless undertaking with the result that the finest achievements of science are pushing man toward the brink of self-destruction. Therefore, as Born writes, "We should resume contact with Goethe and with those who keep cultivating and developing his thought," and especially "learn from them not to forget the meaning of the whole amidst the fascination of details."[6]

The indispensability of both details and of the whole constitutes also a principal theme in the reflections of Carl Friedrich von Weizsäcker on the cultural significance of Goethe's color theory. His ideas on this point can be found in his postscript to a recent edition of Goethe's scientific writings,[7] in his acceptance speech of the Goethe Medal in Frankfurt in 1958,[8] and in his foreword to

Heimendahl's work.[9] Weizsäcker felt that a serious reflection on the conceptual framework of modern physics might open up possibilities to recover the primacy of wholeness, to develop a consistent type of language free from the source of much cultural evil, the Cartesian cleavage. According to him the dialogue between Goethe and modern physics was possible because both rested on a common ground which Weizsäcker epitomized in the phrase: "Plato and the senses."[10] For Weizsäcker the richness of the Platonic "idea" led to its subsequent bifurcation into form and concept, symbol and law, the unique and the general. Neither component of these pairs does, however, suffice. In one's grasping of the uniqueness of a particular experience it is man's urge for immediate wholeness that dominates. When appraised analytically things are divided, atomized, in order that a general statement may be formulated. The immediate wholeness is perceived as form (Gestalt), a category that is closely akin to Goethe's Urphenomenon, which constitutes the backbone of his philosophy and methodology of science. Goethe's timeliness, Weizsäcker argues, should now be evident. As the threat of atomic annihilation and global hunger are in a sense the consequences of man's weakened appreciation for wholeness and relatedness, it is highly imperative to probe into the deepest layers of Goethe's thought on science and especially on the most exact form of it, physics. It is to this task that we should now address ourselves.

To anyone familiar with both physics and literature it may be a source of giddiness to read what the great poet told Eckermann four years before his death: "As for what I have done as a poet I take no pride in it whatever. . . . But that in my century I am the only person who knows the truth in the difficult science of colors—of that, I say, I am not a little proud, and here I have a consciousness of superiority to many."[11] Surprising as it may be, this startling statement was not the effect of momentary impulse. Several years earlier, on December 30, 1823, he had already emphasized to Eckermann that for the previous 20 years he had been the only one to perceive that Newton and his fellow physicists were in decisive error about the true nature of colors.[12] Convinced of his epoch-making achievement, Goethe brushed aside all criticism for the rest of his life. His guiding principle remained what he had written in 1811 in the wake of the adverse reaction of physicists to the publication of the *Farbenlehre*: "These gentlemen [the physicists] may act as they wish, they shall not in the least eliminate this book from the history of physics."[13]

Actually, Goethe could have referred in his remarks to Eckermann not to 20 but to 40 if not 50 years. From his student days Goethe was under the influence of factors that gradually built up in him a frame of mind unreceptive to the conceptual world of Newtonian or classical physics and in particular to its discussion of colors. First, there was his encounter during his university years with Baron d'Holbach's *Système de la nature*, the perfect embodiment of the nightmare of a consistently "scientific" world view. Its reading left Goethe with a distinct revulsion against exact science as can be seen from his vivid account of his experience postdating it by some 40 years.[14] The exploitation of science by d'Holbach's scientism prompted the young Goethe to throw himself, to quote his words, "into living knowledge, experience, action, and poetising, with all the more liveliness and passion."[15]

Unfortunately, it was not only poetry and social life that served for Goethe as a source of "living" knowledge. He found it also in alchemical writings, a dark field that hardly whets one's appetite for the clarity of mathematical physics. In addition, there was at work in young Goethe a secret frustration. Deep in his heart he wanted to excel as a painter but obviously he lacked talent. Gifted with a unique creative power in the field of poetry he felt no urge to write its theory. But he had to theorize about paintings and colors because, as he put it, he wanted to "fill up the deficiencies of nature by reason and insight."[16]

The moment for the release of his pent-up frustration came as he set foot on Italian soil. The resplendent colors of the landscape impressed him to the very core of his being, to say nothing of the impact made on him by the colorful canvasses of Italian painters. While in Rome he even persuaded a friend, an amateur painter, to try out his budding theories on colors. But the crucial event in the life of Goethe, the physicist, occurred on a day in January, 1780, when he was requested to return a borrowed prism that for months lay unpacked on his desk. On the spur of the moment, he decided to make a last-minute experiment with it. Contrary to his interpretation of the laws of Newtonian optics the whole wall facing the window did not turn into the hues of the rainbow. So he decided that Newton was wrong. It appeared to him that an insurmountable barrier, the unquestioned supremacy of Newtonian optics, had been removed from his way, as if by magic. As he recorded the historic moment in the *Farbenlehre,* "I instinctively exclaimed Newton's theory is false."[17]

In the same passage Goethe also admits that he had no expertise in physical science and that he felt the need of professional advice. But when it was offered to him he turned deaf ears on the friendly and competent words of Georg Christoph Lichtenberg, professor of physics at Göttingen. Goethe's few valuable insights in botany and comparative anatomy only strengthened his self-confidence as a physicist. He did not sense that something was wrong with his contentions when his two short "Contributions to Optics," published in 1791 and 1792, received a unanimously negative reaction on the part of physicists. The scientific journals treated him, so he claimed, with "haughty condescendence. . . . They reported my effort in such a way as to help it sink into oblivion forever."[18]

He decided to be his own guide in physics. When during 1792-93 he had to partake in the campaign against the French, he carried along Johann S. T. Gehler's four-volume dictionary on physics. When asked by Prince Reuss during the bombardment of Verdun what he had in mind, the great poet, according to his own account, "began to speak with great animation of the doctrine of colors."[19] It was with the same animation that he tackled the study of Newton's *Opticks* once the campaign was over. But it was not the spirit of docility that led him, the tyro, through the pages of that extraordinary book. He aimed at its wholesale refutation. The literary result was the Polemical Part of the *Farbenlehre* which is characteristically enough left out time and again from most editions of his collected works though, without reading it, one cannot get a real feeling of the psychological abyss of Goethe's self-deceit.

In that Polemical Part, Goethe charged Newton with obdurate resistance to the light of evidence (§§7, 230, 360, 553),[20] labeled him a Sophist (§582) and a bandleader of Cossacks (§178). He dismissed Newton's experiments as inaccurate and his formulas as inapplicable to facts (§10), characterized his method as a systematic evasion of the evidence (§96). He described Newton's theory of colors as a "pleasant tale" and an "empty illusion" (§205), his accomplishments in optics as "hocus-pocus" (§45), a "rubbish of words" (§635) and the worst example of shamelessness in the history of sciences (§360).

Physicists he spoke of as Newton's herd (§654). The mere

thought of their existence turned him angry. He decried their obstinacy (§211), their eagerness to "cement, patch-up, and glue together, as witchdoctors do, the Newtonian doctrine, so that it could, as an embalmed corpse, preside in the style of ancient Egyptians, at the drinking bouts of physicists" (§471). He wished that Newton's followers should "wear special garments so that they could be distinguished from sane people" (§572). Sadly registering that "the world had believed for a hundred years this trickery of theirs" [Newtonian optics] (§113), he felt his urge justified "to tell all possible evil about them and their originator" (§675). His feeling of righteousness in this regard was so overwhelming that 20 years later when a publisher wanted to drop the Polemical Part from a planned edition of his collected works, Goethe most resolutely defended his vilification of Newton.[21]

In line with his overriding ambition Goethe wanted to know all that had been stated on colors from the pre-Socratics down to his time. The result of his voracious reading was presented as the Historical Part of the *Farbenlehre*. As could be expected, Goethe dwelt at great length on the ancients, the medievals and many a second-rate seventeenth- and eighteenth-century cultivator of optics and on their qualitative if not philosophical statements about light and colors. His attention, however, petered out as he approached the closing decades of the eighteenth century that saw physical optics grow more exact and robust. Obviously he could not bear the evidence mounting against him. It was these prejudices that vitiated the purpose of Goethe's remarkable observations of color-effects that comprise the first or Didactic Part of the *Farbenlehre*. It was offered by him as the new and only reliable form of the study of colors. In it mathematics could have no part whatever. After all, Goethe's main pride was to have demonstrated that a physics without mathematics was possible and indispensable. He never wavered in this conviction of his since he intimated it in his first paper published in 1791. Two decades later he recalled that one of the main reasons for its poor reception was that "nobody had any longer the faintest idea that a physics can exist independently from mathematics."[22]

He correctly diagnosed the consensus. No less correct were all the physicists who found themselves unable to reconcile the contradictions in that section of the *Farbenlehre* where Goethe states his views on the use of mathematics in physical science.[23] His was a frantic fear that mathematics destroyed the beauty and immediacy of observation. Observation meant for him an intuitive, poetic glance, or to use his favorite word, *aperçu*. He believed that only a reliance on *aperçus* unfettered by mathematics can discover and keep in view the form of things through which he believed their ultimate nature was shining forth. It was on these "ultimate natures" or Urphenomena that according to him rested the structure of nature whose dynamism was in turn regulated by various pairs of polarities. White light for Goethe was an Urphenomenon that should have never been analyzed into alleged parts. Colors were not parts of white light but the result of the interplay of two polarities, light and darkness. Moreover, he defined darkness not as an absence of light, but a positive entity. Any not perfectly translucent medium contained in itself a share of darkness, and once light touched upon this "troubled medium" colors were produced.

It was this primitive if not obscurantist conceptual system that constituted the framework of the new physical optics (study of colors) as legislated by Goethe. Actually his bunglings merely showed that if there was any scientific way in which the beauty of nature could not be saved it was along the lines of the *Farbenlehre*. He could not even account for the Urphenomenon of colors, the rainbow. Its explanation along his own optics eluded Goethe all his life, though even during his last months he still sent "solutions" of it to his friend, Sulpice Boisserée, director of the Observatory of Munich.

Boisserée's reply was disappointing. He also knew that the previous 17 years had not witnessed any encouragement for the hope expressed to him by Goethe in 1815, five years after the publication of the *Farbenlehre*: "It will take fifty years before the *Farbenlehre* will be accepted; it is now only for the young unbiased men, with the others there is nothing to do, they sit up to their necks in their system."[24] Perhaps some physicists stuck to their system too rigidly. But they had the right to do so. Their system was a consistent one because it was limited. Of this limitedness their awareness rarely weakened. They were also aware that, contrary to Goethe's claim, neither physics nor physicists were to be blamed for the erosion of man's confidence in the qualitative aspects of the world. The culprit was a philosophy heedless of the limitations of physics and of its own premises. Only with these in mind can one understand the reaction shown ever since by physicists to the *Farbenlehre.*[25]

What Goethe wrote about the reception of his two early papers was also in store for the *Farbenlehre.* Among its supporters "there was not a single physicist."[26] German physicists who took up the subject in the 1810s and 1820s could at most commend the material of the Historical Part. For the "physics" of the *Farbenlehre* they had no time to waste. As one of them, H. W. Brandes of the University of Leipzig, noted, Newton's *Opticks* needed as little defense as did the Copernican arrangement of planets or the inverse square law of gravitation.[27] On occasion their remarks turned the table on Goethe when, for instance, F. T. Poselger took the poet to task, that for all his emphasis on painting and painters, he failed to do justice to the greatest of them all—the Sun. "The Sun," Poselger wrote, "appears to us more Newtonian in mentality than Mr. Goethe would have it. And Newton was really the man to ferret out his artistic technique."[28] For all the personal touch in such comments, the physicists never paid back Goethe in kind for his invectives against Newton and his colleagues. E. L. Malus, the first major French physicist to discuss the *Farbenlehre,* displayed good psychology in noting that Goethe was not likely to make many converts precisely because of his excessive intolerance. "As he condemns indiscriminately all statements of the science of optics, it is not in his book that one should dare search for errors that Newton might have committed."[29]

Among British physicists the first to analyze in depth the *Farbenlehre* was Thomas Young. In addition to his historic demonstration of the wave-character of light, Young did pioneering investigations in the field of physiological optics and he had a thorough familiarity with the history of optics as well. To him the Historical Part of the *Farbenlehre* appeared to exhibit "some industry but little talent, and less judgment."[30] He also carried out an experiment, described as "crucial" by Goethe against Newton, but he observed the opposite to what the poet had claimed to take place. Young also emphasized that Goethe's procedure presented a graver threat to the reality of colors than did the Newtonian approach. But most significantly Young called attention to the cultural debacle that would follow if one were to take seriously a work whose demonstrative value was equal to that of the "almanack of muses [full] with epigrams and satires."[31]

It was only a quarter of a century later, in 1840, that another

British physicist of stature took up the matter. David Brewster, who earned his fame mainly by his contributions to optics, joined the battle following the translation into English of the Didactic Part of the *Farbenlehre*.[32] He saw a major threat arising from it, not so much to the "edifice of Inductive Philosophy," but rather to culture in general. He was in particular concerned by the claim expressed by the translator and shared by those admirers of Goethe who admitted the indefensibility of the physics of the *Farbenlehre*, that "the experiments and views of Goethe are more applicable to the theory and practice of painting than the doctrines of Newton and his followers."[33] To Brewster, acceptance of such a claim was tantamount to a cultural disaster, to "placing the principles of art in direct alliance with error."[34] With these words he touched on the heart of the matter. To admit the scientific nullity of the *Farbenlehre* as a theory of optics, and to claim it as the true theory of art, was not only a strategy contrary to Goethe's intentions. It also meant giving respectability to a mentality for which "the slightest resemblances, the most fortuitous associations, are linked together as cause and effect." But as Brewster bluntly put it, it was the sacred duty of scientific thinking to protect culture from the mirage of what he aptly called cabalistic formulas.[35]

The next important physicist to weigh the merits of the *Farbenlehre* was Hermann von Helmholtz. His two discussions of the matter separated by a span of 40 years bring us face to face with a specifically German aspect of the problem. It seems that since the resurgence of German national pride in the 1840s a vindication of the honor of the author of the *Farbenlehre* has become a sacred cause to which German physicists feel duty bound to volunteer their stature and eloquence. Not that German patriotism could derive much comfort from Helmholtz's first address delivered on January 18, 1852, then Coronation Day in Prussia, before the Königsberg branch of the prestigious Deutsche Gesellschaft. As the lecture is widely available in English translation[36] no summary will be made here of Helmholtz's devastating criticism of Goethe, the physicist. The concluding part of the address which bears on the cultural relevance of Goethe the scientist should, however, be singled out. According to Helmholtz, Goethe's devotion to ideal beauty and culture had a very serious shortcoming. It led him to disregard a part of reality, the reality of the backstage consisting of levers, cords, and pulleys. That the sight of the backstage was ugly as compared to the colorful glitter of the artistic performance itself, Helmholtz readily admitted. Yet the machinery of the backstage, or the machinery of the physical world, constituted a reality that man could ignore only at the price of bringing on himself the collapse of civilization. Helmholtz tried to impress this on the humanistic segment of his audience in words that are worth being quoted in full: "We cannot triumph over the machinery of matter by ignoring it; we can triumph over it only by subordinating it to the aims of our moral intelligence. We must familiarize ourselves with its levers and pulleys, fatal though it be to poetic contemplation, in order to be able to govern them after our own will, and therein lies the complete justification of physical investigation, and its vast importance for the advance of human civilization."[37]

Forty years later Helmholtz's tone was notably different. In an address delivered in Weimar at the invitation of the Goethe Society, Helmholtz dwelt at length on what is common between the aspirations of the scientist and the artist. This was unquestionably a skillful approach, through which one could extol Goethe as an apostle of culture and still be left with room to vindicate physics. To develop the former theme was Helmholtz's chief concern, and it led him astray on several points concerning the history of physics. Thus he claimed that the latest developments in the conceptual foundations of physics indicated a rapprochement toward Goethe's preference of the "concrete." As an example, he referred to the growing disfavor in physics for such "abstract" notions as force. He attributed Faraday's success in physics to his sense of the concrete unspoiled by mathematical abstractions. Helmholtz also felt that Huygens' wave theory of light as opposed to the corpuscular (Newtonian) theory was much akin to the Goethean Urphenomenon. Evidently, it was impossible for Helmholtz to finish his address without asserting himself as a physicist. As his lecture drew to an end, Helmholtz restated the indispensability of the inductive, quantitative method of physics and dismissed Goethe, the physicist. "In areas where only the use of inductive method could have helped him, Goethe ran aground."[38]

And so Helmholtz stopped short of doing precisely that against which a distinguished British physicist, A. Schuster, had warned a few years earlier in speaking about Goethe: "We shall not render his memory a service if we convert our admiration for him into idolatry, and bend our knees to his foibles as well as to his strength."[39] In spite of its shortness, Schuster's paper, a true gem in the vast literature on the *Farbenlehre*, brings out more lucidly than many longer discussions the crucial reasons for Goethe's error. The *Farbenlehre* renounced the quantitative analysis of light and colors while claiming to offer a *full* treatment of colors. Another point Schuster emphasized was Goethe's uncritical trust in the reality of the commonsense world. This attitude of Goethe harked back to some basic principles of Aristotle's organismic physics, and the *Farbenlehre* was in more than one sense an effort to turn the clock of scientific history back by several hundred years. Needless to say, such a step can hardly be considered as culturally beneficial, though as Schuster aptly noted, there were still many admirers of Goethe who thought that in the *Farbenlehre* he "may have sown a seed which will bring forth good fruit hereafter."[40]

Possibly such a hope animated Thomas Carlyle. In May, 1878, he had presented John Tyndall, Faraday's successor at the Royal Institution, with a copy of the original edition of the *Farbenlehre*. It was a gift of Goethe to Carlyle, the rising young literary critic who did so much to spread the poet's fame abroad. Carlyle was now speaking of his wish that Tyndall, the noted physicist, examine the volume and set forth what it really contained. Tyndall obliged by graciously taking a second look at the *Farbenlehre*. Following an earlier perusal of the work he had laid it aside with the conclusion that "Goethe in the *Farbenlehre* was wrong in his intellectual and perverse in his moral judgments."[41] His second study of it presented him with no reason to change his first conclusion. In a lecture delivered at the Royal Institution in March, 1880, Tyndall gave a detailed account of the principal points of Goethe's color theory, of its main errors, and of its merits as well. Some of these were cultural but, as Tyndall emphasized it, of rather limited value. As he correctly diagnosed the case, Goethe failed to understand the spirit of tolerance and pluralism which nature is capable of exhibiting when she inspires both poets and scientists.[42]

This spirit of tolerance, the readiness to recognize the irreducible plurality of aspects that make up the fullness of man's search for understanding, was not Goethe's strong side. As Arthur Sommerfeld noted in a thoughtful essay,[43] Goethe was dominated by an irresistible proclivity to take for real only that which was beautiful. Though undoubtedly a great friend and a most sympathetic

observer of nature, Goethe did not possess the temper and qualities
of a systematic investigator of nature. A similar point was emphasized a few years later by another prominent physicist at the University of Munich, Wilhelm Wien, a Nobel laureate.[44] According to
him, Goethe, imbued with pantheism, felt confident that his genius
could establish an intuitive contact with the world-spirit and gain
thereby an all-embracing knowledge of nature. Thus while Goethe
emphasized the observation of nature, he rejected systematic,
laborious, carefully controlled experimentation about the details of
nature. Such was a stance diametrically opposite to the spirit of
physical science and Wien felt impelled to state: "High as Goethe
stands in the esteem of physicists, the dictates of the goddess [the
spirit of science] stand higher than those of even the most important mortal."[45]

But important he was—culturally, that is. And so Wien with a
baffling inconsistency went on to finish his lecture with assertions
on the cultural significance of Goethe in an age of science. It was a
strange effort on the part of Wien who in the same lecture minced
no words in deploring Goethe's inattention to the limitations of
human intellect, however brilliant, and to man's inability to wrest
the secrets of nature through one single masterstroke. Wien could
hardly sound convincing to anyone who had already agreed with his
carefully documented remarks on Goethe's ghastly misconceptions
about physics and scientific method. Yet Wien claimed that some of
Goethe's remarks on the relation of physics and philosophy were
truly remarkable. He claimed that Goethe's Urphenomena and the
fundamental laws and concepts of physics were akin in character.
It was ironically fitting that he should offer a truly enigmatic illustration to a puzzling claim. In a staggering remark about magnetism, which Goethe considered an Urphenomenon, Wien stated
that although electricity gives rise to magnetism, the latter is different in "nature" from the former. Again, one wonders if
Goethe's cultural stature was enhanced by Wien's assertion that
Goethe was not altogether wrong about the sometimes pathological
nature of experimental physics, as even theoretical physics was not
without some pathological aspects of its own. Finally, and contrary
to Wien's claim, physicists did not need Goethe's faith in nature
and in man to press on with confidence in their arduous search for
the laws of nature. They did their work and are still doing it regard-

less of Goethe and his alleged significance for a cultural unity in an age of science.

As proof of this one need only read Weizsäcker's, Heitler's, Born's, and Heisenberg's words on Goethe. On one specific point they all are in agreement: Goethe's dabbling in physics is not physics and under no circumstance should physics consult Goethe the physicist. As to their contention about the cultural relevance of Goethe in the atomic age, one could only wish that they had in mind the real Goethe and not an abstraction of him. The real Goethe did not understand the true nature of exact sciences and their crucial role in modern culture. He was much more at home in the sentimental estheticism of Weimar than in the cruelly muscular world of rising industrialism. To be sure, neither of these milieux represented a healthy solution. The former could not be maintained for too long; the latter could not be escaped any longer. But this is what Goethe tried to do, and in his frustration he laid the blame on Newtonian physics for cultural diseases for which the responsibility lay not in science but in scientism. To overcome the reductionist atmosphere of scientism he offered a reductionism in reverse in which esthetics dominated everything. He failed to see that the tension between the realm of qualities and quantities is an irreducible one which shall not disappear by shortchanging one realm or the other. Esthetical reductionism had few more persuasive if not seductive spokesmen than Goethe. But precisely because of this, in an age so dependent on the proper and unreserved use of science, Goethe cannot be looked upon as a reliable guide toward a truly human culture wiling to render its due to quantities as well as qualities.

[1]W. Heisenberg, *Philosophic Problems of Nuclear Science,* translated by F. C. Hayes (London: Faber & Faber, 1952), pp. 60-76.
[2]W. Heitler, *Man and Science,* translated by Robert Schlapp (New York: Basic Books, Inc., 1963).
[3]Ibid., pp. 24-25.
[4]M. Born, "Betrachtungen zur Farbenlehre," *Naturwissenschaften* 50 (1963), pp. 29-39.
[5]E. Heimendahl, *Licht und Farbe: Ordnung und Funktion der Farbewelt* (Berlin: Walter de Gruyter, 1961).
[6]See "Betrachtungen zur Farbenlehre," p. 39.
[7]C. F. von Weizsäcker, "Nachwort" in *Goethes Werke* (Hamburg: Wegner, 1948-1960), Vol. 13, pp. 537-554.

[8]C. F. von Weizsäcker, "Goethe und die Natur," *Gegenwart* 13 (Sept. 6, 1958), pp. 555-57.

[9]*Licht und Farbe*, pp. vii-x.

[10]"Nachwort," p. 538.

[11]J. P. Eckermann, *Gespräche mit Goethe in den letzten Jahren seines Lebens*, H. H. Houben, Ed. (Leipzig: F. A. Brockhaus, 1925), p. 261.

[12]Ibid., p. 426.

[13]In a letter to F. A. Wolf, 28 Sept. 1811, in *Goethes Sämtliche Werke* (Munich: Georg Müller Verlag, 1909-1932), Vol. 23, p. 129.

[14]J. W. von Goethe, *Truth and Fiction Relating to My Life*, translated by John Oxenford (London: John C. Nimmo Ltd., 1903), Vol. 2, pp. 108-110.

[15]Ibid., p. 110.

[16]"Konfession des Verfassers," *Zur Farbenlehre. Historischer Teil* in *Goethes Sämtliche Werke*, Vol. 22, p. 379.

[17]Ibid., Vol. 22, p. 384.

[18]Ibid., Vol. 22, p. 389.

[19]J. W. von Goethe, *Kampagne in Frankreich 1792*, Ibid., Vol. 34, p. 210.

[20]The numbers in parentheses refer to the paragraphs of the Polemical Part of the *Farbenlehre* as numbered by Goethe.

[21]*Gespräche mit Goethe*, p. 396.

[22]*Goethes Sämtliche Werke*, Vol. 22, p. 389.

[23]Ibid., Vol. 21, pp. 184-186.

[24]*Goethes Gespräche. Erster Teil*, in *Gedankausgabe der Werke, Briefe und Gespräche* (Zurich: Artemis Verlag, 1948-1954), Vol. 22, p. 797.

[25]There is an immense literature on Goethe's *Farbenlehre* listed in the standard Goethe bibliographies. In this article the discussion is confined to comments made on it by physicists of some stature.

[26]*Goethes Sämtliche Werke*, Vol. 22, p. 387.

[27]H. W. Brandes, "Farbe," in *Johann Samuel Traugott Gehler's Physikalisches Wörterbuch*, new edition by Brandes, Gmelin, Horner, and others (Leipzig: E. B. Schmickert, 1825-1845), Vol. 4, Part 1, pp. 39-131; see especially p. 67.

[28]F. T. Poselger, "Der farbige Rand eines durch ein biconvexes Glas entstehenden Bildes, untersucht, mit Bezug auf Herrn von Goethes Werk *Zur Farbenlehre*," *Annalen der Physik* 37 (1811), p. 154.

[29][E. L. Malus], "Traité des couleurs par M. Goethe," *Annales de chimie* 71 (1811), pp. 199-219; quotation on pp. 218-219.

[30][T. Young], "*Zur Farbenlehre*. On the Doctrine of Colours," *Quarterly Review* 10 (1814), pp. 427-441; quotation on p. 429.

[31]Ibid., p. 428.

[32]*Goethe's Theory of Colours*, translated by Charles Lock Eastlake (London, 1840).

[33][D. Brewster], "Goethe's Theory of Colours," *The Edinburgh Review* 72 (1840), pp. 99-131; quotation on p. 99.

[34]Ibid., p. 100.

[35]Ibid., p. 122.

[36]H. von Helmholtz, "On Goethe's Scientific Researches," in his *Popular Scientific Lectures,* selected and edited with an Introduction by M. Kline (New York: Dover Publications, Inc., 1962), pp. 1-21.

[37]Ibid., p. 20.

[38]H. von Helmholtz, "Goethes Vorahnungen kommender naturwissenschaftlicher Ideen," *Deutsche Rundschau* 72 (July-Sept. 1892), pp. 115-32; quotation on p. 132.

[39]A. Schuster, "Goethe's *Farbenlehre,*" in *Publications of the English Goethe Society, No. 5, Original Papers* (London: David Nutt, 1890), pp. 141-151; quotation on p. 141.

[40]Ibid., p. 141.

[41]J. Tyndall, "Goethe's *Farbenlehre—(Theory of Colours),*" *The Popular Science Monthly* 17 (1880), pp. 215-24, 312-21; quotation on p. 215.

[42]Ibid., p. 321.

[43]A. Sommerfeld, "Goethes Farbenlehre im Urteile der Zeit," *Deutsche Revue* 42 (1917), pp. 100-107. Sommerfeld gives an appreciative account of Goethe's original observations about color effects. Characteristically enough Sommerfeld's paper was prompted by his anxiety lest a group of painters in Munich should mislead the Ministry of Education by their insistence on the superiority of Goethe as "color-physist" over Helmholtz, who, according to the group, did not know "light-energetics."

[44]W. Wien, "Goethe und die Physik," in Wilhelm Wien, *Aus dem Leben und Wirken eines Physikers* (Leipzig: Ambrosius Barth, 1930), pp. 79-102. Wien's lecture was given on 9 May 1923.

[45]Ibid., p. 87.

6

A Hundred Years of Two Cultures

A little over a hundred years ago a lecture was delivered at the South London Working Men's College on the topic, "A Liberal Education; and Where to Find It."[1] The renown and personality of the speaker, Thomas Henry Huxley, guaranteed that the audience would be treated to some fireworks. Ever since Huxley vanquished Bishop Wilberforce over the merits of Darwin's theory of evolution, Huxley's fame as a debater became a legend in England. And debate he did. "Where there was strife, there was Huxley," recalled a contemporary wit,[2] and in the late 1860s few issues caused more strife in England than the question of how to update British education.

Huxley was, of course, far more than a debater spoiling for strife. He grasped much better than Darwin certain philosophical issues about evolution and he could crystallize them with stunning con-

This and the next chapter represent the text of lectures delivered on February 26 and 28, 1975, to inaugurate a lecture series on culture and knowledge sponsored by Assumption University, University of Windsor. Reprinted with permission from *The University of Windsor Review* 11 (1975), pp. 55-79 and 80-103.

creteness. His lecture became the focal point of the debate which suddenly grew sharp at least in England between scientists and humanists. Elsewhere the debate had already produced some memorable remarks and created widespread repercussions. Long before Huxley, Pascal spoke of the deep difference between the geometrical mind and the intuitive spirit.[3] A century later d'Alembert took to task those who felt no embarrassment when faced with a choice between Newton and Corneille.[4] But shortly afterwards the "prophets of Paris" launched the age of social engineering based exclusively on Newtonian science,[5] while Germany, largely under the influence of Goethe and the Idealists, fell prey to the Romantic movement.[6] One of its chief aims was to save man from the clutches of a "soulless" mechanistic science. In England it was not until the second half of the nineteenth century that the same tension between the sciences and the humanities came to a head.

The combatants, to recall Huxley's word, formed two camps, both with firm convictions about what culture ought to be. Between those two camps lay the mist of mutual incomprehension and mistrust. The latter had much to do with the rapid shift of power from the men of letters to the men of science. In 1832, the second meeting of the British Association for the Advancement of Science ran into spirited protests on the part of some Oxonian teachers of the humanities. The famous Keble argued that scientists were "a hodge-podge of philosophers"[7] unfit for honorary degrees from Oxford. A generation later the annual meetings of the Association commanded world-wide attention. Confident of riding the wave of the future, the Association urged in 1868 that courses in science be given a substantial role in British secondary education. In doing so the Association asked for nothing less than for laying the foundation of a new cultural orientation. When in the same year Huxley told the fledgling Working Men's College that it should never imitate, even if it could, Oxford and Cambridge, the paragons of British higher education, he cast in effect his vote for the replacement of one culture with another. What is more important, he did so in a manner that galvanized the attention of his generation.

As an accomplished strategist, Huxley considered attack the best defense, and promptly carried the fight into his opponents' camp. Weak points, defenseless positions he could find there by the score. It was hardly a feat for him to show that the so-called traditional or literary education as practiced in most English schools of his day was at times a barren enterprise. Yet nothing he said about the glaring shortcomings of the curricula of elementary and secondary schools, and even of universities, constituted a revelation. For more than a decade the British public followed closely the work of various government commissions that tried to inject new vigor into British schools at all levels. For all that, Huxley's galloping over a fossil-looking enemy could not fail to catch public imagination. The image Huxley wanted to paint was not that of a temporary campaign. He was more concerned with the vision of a mankind permanently steeped in culture to which the spirit and findings of science provided the foundations.

By foundations Huxley meant much of the superstructure as well. Moreover, the whole edifice of culture as pictured by him looked more like a factory than a dwelling place. Its occupants were not so much individuals as specimens developed through a complete conformity to the laws of nature. As one would expect, these specimens excelled by their machine-like characteristics. Their intellect was described by Huxley as a "clear, cold, logic engine, with all its parts of equal strength, and in smooth working order, ready like a steam engine."[8] In the same vein he characterized their emotional and moral behavior as the quiet hum of a system perfectly tuned to its environment. Although Huxley also spoke in the same breath of "tender conscience" and of "vigorous will," he thought of them in the framework of the evolutionary, mechanistic outlook of which he was a chief champion. In that outlook will, conscience, ethics, love of beauty, and the like could hardly be more than verbal generalizations of the many small and great contacts, if not conflicts, between the individual specimen and its environment, both social and cosmic.

In such a picture the fate of the individual depended entirely on a successful game between him and the surrounding forces. Huxley paraphrased it as a chess game, in which the world formed the board, the phenomena of nature stood for the pieces, and the laws of nature set all the rules of the game. Though he called it a game, he made no secret of its deadly seriousness. It was as deadly as the struggle for survival and was identical to it. In that game or struggle there was no room for aims and goals, only for successive phases of perennial transformations. The truth was even worse than that but it dawned on Huxley only a decade or two later. Yet the truth had already been clearly stated in better textbooks of physics. There Huxley could read, if he wanted to, the three laws of thermodynamics. The first of them states that you cannot win, the second that you cannot break even, the third that you cannot even get out of the game. All this meant that individuals as well as nature were certain losers when viewed only through the laws of nature. Nothing of this was intimated when Huxley spoke in the same breath of nature's laws and of the laws of man's nature and defined the aim of education as the molding of man's inclinations into full compliance with those laws. Culture was the result of such an education.

Huxley had a vibrant and warm personality which could hardly feel at ease with such an account of culture. Its harmony was cold, machine-like at its best, and cruelly remorseless and inexorable at its rawest. It made an adverse impression to say the least, and Huxley did his literary best to accommodate beauty, purpose, and warmth in that desolate picture. He even made ardent protestations against the inference that he was an enemy of classical studies. Had his first opportunities for education been different, he remarked, he himself would have chosen classical philology as his life-long pursuit. Yet while Huxley strove for a catholicity of spirit, he was more firmly locked in the tracks of mechanistic evolution than he suspected. It served him as the scale by which all else had to be sized up, classical philology and history not excepted. If he found a saving grace in them it was only so because, as he put it, these two subjects, if properly pursued, were akin to investigations in paleontology.[9] In other words, both were equally useful and noble preoccupations provided they illustrated Darwinian theory which he took for the general and basic explanation of man and the universe.

In that theory Huxley saw the highest vision man could ever aspire to. In it he could find no fault, no shortcomings. It represented to him the best, the most enlightening which the ideology of science could provide. Nor did he see any fault with science. He saw it, its method, and cultivation, and even its cultivators, in their ideal form, under the most favorable light. Engrossed with the ideal notion of science and science-based culture, he contrasted it with literary or traditional education as given in the classrooms of his own time. One wonders if his tactic earned him more than a Pyrrhic victory. After all, nothing lasting has ever come out of one's eagerness to capitalize on a comparison in which the *actual* imperfection of one side is set against the *ideal* perfection of the other. Such a procedure is equivalent to relying on a heavily tilted balance, a point that can easily be overlooked by any over-committed mind. Huxley's was certainly such a mind, for all its brilliance and good intentions. No wonder that he also fell short of the highest standards when he tried to give explicit attention to the inherent merits of literary education or culture.

This he did twelve years later when he journeyed to Birmingham to deliver a lecture on "Science and Culture"[10] in a newly founded college, which offered mainly courses in science and technology. The new school owed its existence to the munificence of Sir Josiah Mason, who stipulated that neither party politics, nor religion, nor "mere literary instruction and education"[11] should be pursued within its walls. The barring of the first two subjects appeared to Huxley an unmitigated blessing, but he felt that the provision on "mere literary instruction and education" called for some explanation. Obviously, he could not give it without broaching the general question of culture. Much less could he discourse on culture in general without referring to what was said about it by that "chief apostle of culture,"[12] as he called his good friend, Matthew Arnold. In fact, he centered his whole lecture on a statement of Arnold in which culture was defined as the endeavor "to know the best that has been thought and said in the world." The phrase came from Arnold's famous essay on "The Function of Criticism at the Present Time" (1864),[13] in which Arnold also stated that culture was identical with the criticism of life contained in literature.

With the first of Arnold's statements on culture Huxley had no quarrel. He agreed that culture was more than technical skill, that it implied "the possession of an ideal, and the habit of critically estimating the value of things by comparison with a theoretic standard. Perfect culture should supply a complete theory of life, based upon a clear knowledge alike of its possibilities and of its limitations."[14] This was an admirable and unexceptional statement, as it implied the acceptance of the principle of universality as the uppermost standard in man's search for truth and values. When it came, however, to its being translated into practical applications Huxley's fighting spirit considerably narrowed that most valuable standard. That fighting spirit made him read into Arnold's second statement a narrow notion of literary pursuit, and he picked it to pieces with all the impatience of a crusader. Thus he declared that it was unthinkable for him to consider literature as the competent source of the essence of culture, "the criticism of life." To prove his point he took again the facile way. It consisted in dwelling on the malpractice of literary men, just as he rested his case twelve years earlier on the malpractice of literary educators. Once more he tried to heap indirect accolades on science by attributing insensitivity toward science on the part of men of letters.

It was a clever tactic not to antagonize the living if the goal could be achieved by taking to task the dead. Before his audience could realize it, Huxley was already painting with quick strokes the image of a Middle Ages lost in Latin grammar and theological hair-splitting. Living in an age that still believed that everything was dark before Galileo and all was sunshine after him, Huxley could hardly guess that many medievals showed far greater interest in ancient Greek science than did most humanists of Renaissance times. More culpable was Huxley in proclaiming his "unhesitating faith that the free employment of reason, in accordance with scientific method, is the sole method of reaching truth."[15] He should have known that the great living figures of British science, Faraday, Maxwell, and Thomson, the future Lord Kelvin, and a host of others, repeatedly and publicly rejected what for Huxley was an article of faith: "the applicability of scientific methods to the search after truth of *all* kinds"[16] (italics added).

Truth for Huxley was either measurable, indirectly at least, or it was not truth at all. It was a position that could prove stifling and Huxley, owing to his natural instincts, felt its strictures keenly at times. It was under such a momentary impulse that he declared that "exclusive scientific training" could bring about a mental twist. Yet he was unable to deplore anything connected with science without indicting in the same breath the opposite side as well. Exclusive literary education was for him just as harmful as a purely scientific one. Not that he wanted to question openly the importance of "genuine literary education." He noted with satisfaction that courses in English, German, and French were provided for in the curriculum of the new college, and he expressed his hope that students would go on studying the respective literatures as well.[17]

Whether they did or not mattered little in the end to Huxley. It was of little consequence to hear from him that "if an Englishman cannot get literary culture out of his Bible, his Shakespeare, his

Milton, neither . . . will the profoundest study of Homer and Sopho-
cles, Virgil and Horace, give it to him."[18] No wonder that in extol-
ling modern, especially English literature, as opposed to classical
lore, Huxley ended up in clear inconsistency. At the very outset of
his lecture he claimed that classical studies justified neither time
nor expenditure. To this he added that a purely scientific education
was at least as good as a purely literary one. He now clearly meant
that the scientific one was far better.

The reason for his inconsistency is not difficult to find. Being
overawed by the so-called scientific method and scientific world-
picture, Huxley could not be attracted to what only introspection
was capable of telling about man. His interest in man was
restricted to propositions that tackled the enigma of man not from
the inside but from the outside. The inside part he mistrusted as
intractable to reason; the outside he was willing to appraise only in
terms of laws that aimed at the exactness of physics. Thus he urged
that man above all must learn that his individual and social attitudes are the expressions of natural laws, the laws of social statics
and dynamics. He had the firm conviction that due to the irresistible march of the scientific spirit man before long would deal with
political questions in the same manner as scientific questions are
handled.[19] The correctness of Huxley's prognostication depended,
of course, on his assertion that man's individual and social actions
represented a machinery as delicate, by which he meant exact, as
that of a spinning-jenny.[20] The granting of this assumption could
only make the role of literature in education and culture more pre-
carious than ever. For literature is not a recital of data of social
statics and dynamics, nor is it a compendium of laws about man
and society verifiable by Pavlovian experiments or by opinion polls.
Literature is rather the mirror of man himself in his actuality and
concreteness. It is a mirror that shows details and conveys shades
of meaning that alone make man's culture fully human.

Such are, however, considerations that hardly impress minds
devoted to that cultivation of science which turns science into
scientism, or the belief that the method and results of the science of
the day have an exclusive competence in all fields of inquiry. To
make matters worse, the number of such minds was growing by
leaps and bounds, so as to threaten literary culture with rapid
extinction. The London *Times* took the view that in another hundred years only a few eccentrics would read literature. On the con-

tinent Renan assigned the same hundred years, ending in the 1970s, as the expiration date of literary, historical, and critical studies. The prospect for the humanities could indeed appear very gloomy to Arnold, himself a poet, who referred both to the *Times* and Renan in the opening section of his Rede-lecture on "Literature and Science"[21] delivered at the University of Cambridge in 1882. Arnold offered his lecture as a programmatic defense of the irreplaceable role of literature in human culture.

To achieve his aim Arnold had several avenues open to him. He could have followed Huxley's strategy by carrying the fight into the opposite camp. Laying bare the shortcomings of scientific pursuit as *actually* practiced would have no doubt been a very effective strategy. The scientific scene of Arnold's day provided more than enough grist for his mill. Then as now the atmosphere of science was made up of a mixture of reliable observations and vague guesses, of sharp insights and unwarranted generalizations, of indispensable metaphysical assumptions and "ersatz" metaphysics. To give an impressive portrayal of this would have required a good inside view of what went on within the scientific realm, but this view Arnold did not possess. Nor was it his intent to make much hay of cases in which competence in science served as the springboard for somersaults into foreign waters. The only example Arnold mentioned was that of a member of the British Parliament, a fine geologist, who traveled all over America, and made an excellent evaluation of the mining prospects in the United States. The final conclusion of his geological survey was that only by becoming a monarchy would the States secure a happy future for themselves. He even specified that the United States should offer the throne to a member of the British Royal family, and that the Senate should be replaced with a House of Lords selected from among the great land-owners. Happily for Lincoln, he was already dead. Less than a hundred years later the American Secretary of State, James Byrnes, was so bitter about some scientists making oracular statements in politics as to declare: "Today every man must have his own physicist."[22]

As to science, Arnold's information was meager. According to his own admission he had to learn, or rather re-learn, elementary arithmetic after he had been made inspector of education. Still Arnold knew that the scientific pursuit aimed at two principal targets: the acquisition of technical skills and the development of concepts and theories of an increasingly more universal character. It was this theoretical part of the scientific enterprise that made unique contributions to man's intellectual grasp of the universe and Arnold would have been the last to underestimate their importance. But in his view knowledge constituted only one of the principal domains of man's aspirations. In addition to the urge to know in a scientific fashion, there was in man an innate devotion to beauty, to moral principles, and to social harmony. Moreover, the development of the full range of man's ability to know could not be accomplished in isolation from man's other capabilities and exigencies. Knowledge implied on an essential basis an inescapable readiness to relate any and all pieces of information to the whole texture of human enterprise, that is, to man's moral, esthetical, and social world. What this meant was that underlying man's purely intellectual capacity was the much broader entity of human nature, and that culture consisted in doing justice to all aspects of that nature.

Once this was made clear, Arnold could easily dispose of the secondary question as to what constituted the true definition of literature as the primary source of "the best which has been thought and said in the world."[23] While Huxley claimed that Arnold meant only *belles-lettres,* Arnold insisted that his idea of literature included Euclid and Newton, as well as the poets. To know the Greeks of old, it was not enough to be familiar with their poets. Their insights into science were equally revealing, and the same held true of the British. Their culture rested on Newton no less than on Shakespeare. Huxley could hardly object to this, openly at least. Their radical disagreement turned around Huxley's contention that for the great bulk of mankind, and especially in modern times, physical science should form the staple of education.[24]

The main trouble with this, Arnold remarked, was that it ran counter to the basic needs of human nature deriving from its full potentialities. First, the great majority of those receiving a preponderantly scientific education would remain on the level of scientific technicians, whose knowledge was a mere "instrument knowledge,"[25] hardly satisfying the deeper urges of man to know. The mentality of such men would, therefore, be stunted with no small detriment to their own cultural attainment and to culture in general. Second, even among scientists working at the more theoretical and creative level there was, as Arnold aptly noted, only one Darwin for fifty Faradays.[26] By this he referred to Dar-

win's disclosure about his own ability to live without poetry and
without the emotional richness implied in its enjoyment. Darwin's
case was not typical though very revealing. It illustrated the insensitivity that can seize a great mind exclusively devoted to a
systematization of the data of the external world. The personality
of Faraday stood for the balance and richness of the inner world of
no less an eminent investigator of nature.

To develop and maintain the vigor of that inner richness man
needed, so Arnold argued, poems, novels, and dramas, or *belles-
lettres* in short. They were as many mirrors of the richness of man's
inner world as the phenomena of nature were the reflection of its
orderliness and rationality. While careful observation of nature
provided the spark for the discovery of its laws, the *belles-lettres*
were meant to exercise a similar sparking effect. Pieces of *belles-
lettres* were designed to enkindle in man an inner chain reaction to
be followed by the activation of his esthetic, moral, and social
energies. Poems, novels, and dramas were particularly suited to
perform this task, because their message was carried in a language
through which such aspects of reality could be intimated and
powerfully evoked, that were inaccessible to scientific discourse. It
was not exactly the same to say that "patience is a virtue," as to
repeat with Homer that "an enduring heart have the destinies
appointed to the children of men."[27] The meanings of the two
phrases were essentially identical, yet the impressions produced by
them had an unmistakably different ring.

According to Arnold's principal thesis the existence of a rich and
healthy human culture depended in a large measure on man's sensitivity to such differences. An almost exclusively scientific education threatened precisely this sensitivity. For illustration Arnold
referred to a case which fully anticipated the problem of two cultures posed in our time by efforts to formalize language, or to com-
puterize it. For the phrase, "can you not wait upon the lunatic?" is
as flat as a punch card and therefore would readily be absorbed by a
computer. But the technical student of Arnold's day was totally
wrong in thinking that the phrase presented an adequate rendering
of the richness of the line from *Macbeth*: "Can'st thou not minister
to a mind diseased?" The student was as wrong as that computer
which rendered a century later the phrase, "out of sight, out of
mind," as "invisible lunatic."

In the great debate between Huxley and Arnold nothing was

more relevant or of greater future significance than Arnold's illustration of that type of meaning which can be conveyed by literature alone. He also touched on the heart of the matter by showing how deeply the exclusive message of *belles-lettres* was tied to those aspects of man's nature that could not be fathomed, let alone activated, by the method and language of science. It is much to be regretted that Arnold's discussion of this point was relatively brief.

Most comments that had been made during the ensuing decades on the respective merits of the two cultures contained little insight and much partisanship. Such was certainly the case about the apostles of scientism preaching the messianic age and its being brought about by man's exclusive reliance on the method and products of science. Their message echoed what Herbert Spencer put down somewhat earlier as the manifesto of educational and cultural reform. The manifesto stated in a rolling if not inflated Spencerian rhetoric that science was the best and most needed tool in every facet of human life, including family relations, political activities, and religious practices. Spencer based this gigantic claim on the contention that the truths of science were "necessary and eternal," and that "all science concerns all mankind at all time."[28]

What all this proved was that Spencer's disagreement with the scientism of Comte's positivism concerned only particular points, not its essence. He embraced it wholeheartedly, together with its ideological background, the social engineering of the Enlightenment. D'Holbach, Condorcet, Saint-Simon, and Fourier deserve special mention in this connection. The cultural program of their scientism was rekindled a century later in some prominent men of science, like W. Ostwald, and given immense popularity in the novels of H. G. Wells. Needless to say, it took a thoroughly reconditioned man to feel at home in the new, scientifically engineered world of *The Time Machine* (1895), of *Mankind in the Making* (1903), and of *A Modern Utopia* (1905). More ominous than the fancy of a novelist was the fact that before long the notion of social engineering made heavy inroads into various branches of the scientific study of man. Behaviorism took psychology and sociology by storm, showing to what degree some scientific circles had already been under the sway of scientism.

Between the two World Wars the cultural atmosphere was filled to the saturation point by statements advocating both the feasibility and desirability of a "retooling" of mankind according to carefully planned scientific specifications. The new "scientific" picture of man stood in sharp contrast to the actually existing man, who was anything but a cold logic machine, free from emotional pecularities, free from often tragic urges, free from whims and fancy, and last but not least, free from hardly definable creative impulses. To replace a man who could love as well as hate with a man conditioned to total conformity meant also the end of a world that literature stood for. One can, therefore, understand that leading men of letters mostly took toward science a negative attitude that varied from lofty disinterestedness to vitriolic hostility, as exemplified respectively by T. S. Eliot and J.-P. Sartre. There was not much balance either in the writings of those who, like Ortega y Gasset, earned during those decades their reputations as philosophers largely on the basis of what they said on culture.[29]

Matters were not helped by the Second World War out of which there emerged the rough and painful structure of a new age. It witnessed not only the redistribution of power and the gigantic confrontation of ideologies. It also brought about the upsurge of new nations and the entry of entire continents into the mainstream of daily history. In more than one sense the world had suddenly become overcrowded with trends, plans, movements, ambitions, intrigues, with a great variety of cultures, and with dire needs. For the first time in history the world had become one in the sense that most political, cultural, social, and economic waves had an amplitude so large as to splash around the entire globe.

Parallel to this development came sweeping transformations in the world of science. Their seeds were germinated, as was the case with the social changes, in wartime tension, secrecy, and confusion. It was then that electronic feedback mechanisms, atomic energy, computer techniques, and automation reached a stage of development where they were no longer mere scientific discoveries but tools of unparalleled power. With proper use they were capable of creating, if not a new heaven, at least a rejuvenated earth, and a new hell if abused or not utilized to their full extent.

All this was abundantly clear by the mid-fifties. There was nothing new in the warning made in 1959 by C. P. Snow, later Lord Snow, that only a resolute application of the possibilities of modern science would spare mankind from a global cultural disaster. As Matthew Arnold did three-quarters of a century earlier Snow, too, used a Rede-lecture as the platform for his concern about culture. His now famous lecture, *The Two Cultures,* had as its ultimate target the deepening abyss between rich and poor nations, and its destructive consequences. Chief among these were an inevitable global conflict, the engulfment of mankind in poverty and disease, and the subsequent disappearance of most if not all forms of cultural refinement.

Of culture he spoke little. He gave no definition of it, as the one he proposed could easily turn into a parody of culture. He saw the heart of culture in the spontaneously uniform reaction of men of science, or in Snow's words: "Without thinking about it, they respond alike. That is what culture means."[30] Such a definition of culture, if not further specified, is fraught with great dangers, because on its basis the Hitlerjugend, the Komsomol, and the Red Guards would equally qualify as cultures and so would a pack of wolves, a bevy of predatory birds, and a swarm of busy-body ants. They all respond alike without thinking about it. Clearly, one needs more than purely behavioristic categories to define culture. Snow gave no convincing evidence that he was ready to transcend those categories repeatedly cut to size long before his Rede-lecture. There was no originality either in Snow's description of our culture as split into two factions, a literary one and a scientific one. Nor did he offer much enlightenment about them either, except some delightful, expressive snapshots concerning their respective practitioners.

It was in these gemlike asides that lay both the attractiveness and the deceptiveness of what Snow presented in *The Two Cultures.* It was not about culture unless one is satisfied with a surface notion of it. Nor lay its chief concern in bridging the two cultures. Apart from some mild protestations to the contrary, he seemed to anticipate joyfully the absorption of literary culture into the scientific. He was more eager to suggest that scientifically trained men displayed far greater and more consistent sensitivity about the "true" problems of their time than did their literary counterparts. He described the latter as born "luddites," or instinctive antagonists of progress, reform, and material improvement. At the same time he claimed that scientifically trained men had always given prompt support to the cause of the future, or to recall Snow's flat dictum, "they had the future in their bones."[31]

Sweeping generalizations hardly ever have deep roots and Snow perfectly illustrated this, as shown by the proofs offered by him to substantiate his contention. According to him men of letters, with the exception of Ibsen, were utterly insensitive to human and social problems created by the Industrial Revolution. The literary history of the nineteenth century—just think of Dickens—is plain enough to give Snow the lie. In a similarly unsound fashion Snow attributed special virtues to scientists as a body. Yet no reliable sociological study has ever demonstrated that scientifically trained men are animated by genuine altruism and that their sense of social justice is always at full vigor. This is, however, what Snow claimed though he seemed to have sensed the incredibility of his assertion. This is perhaps why he poured around it a little reality as he referred to the psychological impoverishment of his heroes, the scientists. Their esthetic grasp, as Snow admitted, hardly extends beyond the realm of music.

The halo assigned to scientists was rather prejudicial in the case of Snow who was eager to note that for several years his main job consisted in commuting between the scientific and literary communities of England. He should have, therefore, had more than one occasion to hear prominent men of science acknowledge that scientists were not any more virtuous or perceptive than members of other professional groups. It is most unlikely that Snow had been unaware of the Presidential Address delivered by A. V. Hill before the British Association for the Advancement of Science at its meeting in 1952. There Hill warned that as regards morality and ethics "there is no such thing as 'the scientific mind'. Scientists for the most part are quite ordinary folk. In their particular scientific jobs they have developed a habit of critical examination, but this does not save them from misrepresentation and falsehood when their emotions and prejudices are strongly enough moved."[32]

Prejudices abound in *The Two Cultures* and form the backbone that holds its four parts together. In the first, scientists are favored over men of letters for their greater social sensitivity. In the second, applied scientists are given the nod over theoretical scientists for the same reason. In the third, those among applied scientists receive Snow's praise who are concerned with the social distribution of the products of technology. Finally, in the fourth, Snow heaps praises on social systems, especially on the Soviet Union, that based their policy of education on training in large numbers this kind of applied scientists. Only one with heavy prejudices about culture could present such sweeping and lop-sided generalizations in the name of culture.

An example of these astonishing generalizations is Snow's discussion of the rise of the Industrial Revolution. It grew robust, so Snow claimed, on both sides of the Atlantic under the guidance of handymen and he listed Henry Ford as their chief example. To call a highly competent engineer a handyman is allowable in poetry, but hardly excusable in a discourse devoted to the virtues of the scientific mind. Those ready to dismiss this remark as pedantry should recall that immediately preceding Snow's portrayal of Henry Ford there lurks another of Snow's hair-raising generalizations which again cannot be traced to sheer ignorance: "So far as there was any thinking in nineteenth-century industry, it was left to cranks and clever workmen."[33] Statements like these refute themselves and one wonders what prompted Snow to believe that his audience would not rise in protest right on the spot. Perhaps he suspected this possibility and to prevent it he relied on the age-old method of throwing a red-herring. He did so by telling a sparkling anecdote. "The academics," which in the context could only refer to professors of science, "had nothing to do with the industrial revolution; as Corrie, the old Master of Jesus, said about trains running into Cambridge on Sunday, 'It is equally displeasing to God and to myself'."[34] By 1959 George Elwes Corrie was truly old, indeed dead for three quarters of a century. Who in Snow's audience knew that Corrie had been Norrisian professor of divinity and not a man of science?

Though slightly misleading, Snow's story is powerfully evocative. It is poetry in prose, with all the psychological impact of the artistic mixing of verbal images (industry, revolution, old Master, Jesus, trains, Cambridge, Sunday, displeasure, God, and myself), a mixing which is the *forte* of literature. Its aim in Snow's handling is not only to illustrate something to the mind, but also to disarm it. The enjoyment of the story then acts as a pill to numb the critical sense of the audience. Once this is achieved anything goes, even outright nonsense. While it has never been profitable to argue against nonsense, it is worth probing into the possible causes that motivate nonsensical statements. One does not have to turn many pages in *The Two Cultures* to find some of these causes or at least one of them. Immediately following the foregoing anecdote about Jesus, trains, old Master, and Cambridge, Snow recalled the fact that in Germany instruction in applied science flourished as early as the 1830s and 1840s, that is, at least a generation earlier than in England and in the United States. To this Snow added a most revealing remark: "I don't begin to understand this: it does not make *social sense*"[35] (italics added). This was an uncanny suggestion that for Snow a culture was worthy of its name if it made social sense, both culture and sense being now reduced to the effectiveness of the tools of production. Indeed, from this point on *The Two Cultures* is not about culture, it is not even about science and literature, but about the organization of society for the reception of the tools of production created by applied science. It is this problem about which Snow declared with grave solemnity: "There never was anything more necessary to comprehend."[36]

It should be of no surprise that Snow evaluated the twentieth-century scientific revolution in terms of its power to produce an entirely new society, the modern "industrial society," which is, to quote him, "in cardinal respects different in kind from any that has gone before." He also submitted that "the personal relations in a productive organization are of the greatest subtlety and interest." From this it followed for him that the true scientific culture is embodied in the applied scientist and engineer. Pure scientists are only an appendix to it as they "have themselves been devastatingly ignorant of productive industry, and many still are."[37]

Such explanations of culture and society in terms of industrial production have a curiously Marxist ring. It should now be easy to understand why of all cultural groups, systems, and ideologies, it is Marxism that usually gets high marks in *The Two Cultures*. Though Snow never used the word Marxist, and only once in a while the word Soviet, it is all too clear what he meant when referring to Russia and the Russians. He could not have Dostoevski, Tolstoy, Pasternak, and Solzhenitsyn in mind when he praised Russian novelists for giving a prominent place to the engineer-type scientist. Snow, of course, kept silent about the literary nullity of poets regimented into singing of tractors and of locomotives. He claimed that the gap between the two cultures in Soviet society seemed not "to be anything like so wide as with us,"[38] while at the same time he ignored that rigid censorship which can only create tragic gaps and is bound to kill anything worthy of the name culture. Snow extolled the U.S.S.R. for training women in large numbers as engineers while he kept mum about the far greater number of women forced to perform manual work suited only for robust men. Again, when Snow praised Russian secondary education as most productive of people who understand culture, he said not a word about the systematic distortion of history, both political and scientific, that runs through Soviet textbooks. Elementary respect for truth was also abdicated in Snow's account of Russian (Marxist) and Western (capitalist) scientific production. He made much of the larger number of engineers trained in Russia than in England and the U.S.A. in tacit compliance with the Marxist dictum that growth in quantity will issue in new quality. He flatly stated that concerning the training of pure scientists in physics and mathematics the balance was heavily tilting in favor of the Russians. Whatever the truth of this evaluation, the fruits of such a heavy imbalance have not yet evidenced themselves in the form of theoretical and technical breakthroughs to be ascribed overwhelmingly to Soviet scientists.

Sensitivity for factual truth was hardly at its best in Snow's silence about some notorious features of a culture which he set up as the ideal for its alleged appreciation of the industrial revolution. Those features were well known facts when Snow spoke. Among them were the opposition of Party ideologues to "bourgeois Einsteinism" in physics, their derision of cybernetics, and their successful destruction of an entire branch of science, genetics. Of Lysenko Snow made no mention. Although he once passingly referred to some "horror,"[39] he did his best to offset its impact on the reader by a little story to the contrary. The story was about John Cockcroft, a Nobel-laureate physicist, who went to Moscow and returned to Cambridge with the news that work was done in the great Moscow factories in much the same sensible way as in the big ones in England. "A fact is a fact is a fact,"[40] was Snow's comment on the story of Cockcroft. Some scientists who did not have the protection of a British passport brought back very different news once they were able to get out. Of these facts, very numerous and widely printed, Snow preferred to keep mum.[41]

Truly, a fact is a fact is a fact . . . but not of course when weighted with Snow's heavily tilted balance. And as questions owe their existence to the recognition of facts, falsely weighed facts give rise to questions whose true weight goes unrecognized. A case in point is when Snow voiced the crucial question without seeing how crucial it was. It can be found in the next-to-last page of *The Two Cultures* where Snow asked: "Can you possibly believe that men will behave as you say they ought to?"[42] Such is the greatest of all great questions and is certainly more important than the question of man's understanding the industrial and scientific revolutions. But this crucial question is flattened in Snow's effort to make it fit the cubbyholes of his covertly Marxist scientism. What seemed crucial to Snow in the question was that while the U.S.S.R. has the political techniques (a curious euphemism for slave labor-camps, suppression of free press, and elimination of free elections) to channel man's capabilities, no effective political techniques are available in the West. Emboldened by the unwisdom of this reasoning, Snow then raised the specter of a West failing both practically and morally.

A truly unexpected sleight of hand. The injection of moral perspectives is anything but logical in Snow's analysis of culture and of its split in two. In resting his case with the primordial importance of the tools of production as supreme factors in human life, individual and social, Snow had as little right to speak about morality, good will, responsibility, and the like as had Huxley to whom culture existed in man's conformity to the laws of nature. Both Snow's and Huxley's views on culture are heavily mechanistic, as is also the case with every definition of culture made on the basis of scientism. In such a conception of culture there remains no logical room for a distinction between tools and goals and for a sensitivity for the utmost importance of this distinction.

That an appraisal of culture in terms of scientism, as given in *The Two Cultures,* would become so popular as it did merely shows the extent to which scientism has eaten itself into the fabric of our thinking. Our alarm is rather subdued on witnessing the stark reality of what for Thoreau was a mere foreboding: "Man has become the tool of his own tools."[43] We do feel but vaguely the dramatic conflict which prompted Captain Ahab in Melville's *Moby Dick* to exclaim in despair: "All my means are sane; my motives and objects are mad."[44] Or if we do, we look for "scientific" ways to cure evils and tragedies caused by an overemphasis on science and technology. Panegyrics of scientism keep reverberating, while it is readily forgotten that prophets of scientism ended up time and again in utter disillusion about its merits. This happened to Huxley toward the end of his life as shown in famous Romanes-lectures on "Evolution and Ethics." It happened to H. G. Wells, as can be seen in his *Mind at the End of Its Tether*. It happened to Bertrand Russell, who in his lectures at Columbia University in 1950 admitted that what the world needed above all was Christian love.[45]

About Christian love one can perhaps say, and saying it one can perhaps hope for general agreement, that Christian love is not a product of science, let alone of scientism. In other words, it is difficult to escape the conclusion that culture has, or rather must have, some crucially important ingredient that cannot be had from science and technology. In short, science with respect to culture is a limited entity. Developing strong awareness of the limitations of science vis-à-vis culture is the pre-condition for taking any positive step toward reuniting the two cultures. The word positive is of special importance in this connection, as not much shall be gained if the cultural apostles of scientism, such as Snow, will primarily be taken to task for not having done their homework in literature. Though this can expose to some degree the shallowness of their thought, it also can make martyrs of them, if in the very same process invectives are resorted to. An illustration is the vituperative language used by F. R. Leavis in his Richmond-lecture, "On the Significance of C. P. Snow,"[46] delivered at Downing College in early 1962. Unfortunately, in his animated analysis of Snow's competence as a scientist, a historian, and a novelist, Leavis devoted only a short section to what is unique in literature.

It was not of Leavis's domain to pass judgment on Snow's familiarity with science. He should have rather taken him to task for his reading of its history. Leavis was more competent in calling attention to Snow's haughty reading of the history of literature. In this connection Leavis scored some points though not too constructive ones. The same is true of Leavis's wasting his authority as literary critic on the lack of literary merit in Snow's novels. Leavis, however, went to the heart of the matter in chastizing Snow for slighting first-rate novels, either because they reflected indifference toward science and industry, or because their authors deserted the working classes.

While for Snow literature had a reason to exist only if it was harnessed in the service of a specific social philosophy, Leavis rightly pointed out that literature best serves society by not serving it. Evidently, the cultivation of literature, in order to fulfill its social function, must remain free of the shackles of sociological categories and fashions. Only in such a freedom can it become the true re-creation of what is actually experienced by man. The vehicle of that re-creation is the language which, as Leavis aptly noted, is prior to any and all science, and its actual cultivation forms the ultimate basis of man's understanding of science in all places and at all times. By this remark Leavis got to the threshold of a goldmine without entering it. Instead, he turned sideways and finished off with some run-of-the-mill remarks on the importance of "living" one's own language and of the art of literary criticism to which he accorded the central role in university education.

The goldmine was left unexplored in that otherwise perceptive analysis of the Snow-Leavis controversy by Lionel Trilling, "Science, Literature, and Culture."[47] In particular, he failed to develop his concluding suggestion that literature was able to disclose areas of the mind that are far beyond the reach of science, which is largely restricted to quantitative, measurable propositions. Comparison of any good poem, novel, or drama with any good scientific paper will readily show that literature and science develop the every-day language, which is the ultimate vehicle of understanding, in two very different directions.

The various steps through which literature produces its specific brand of language were beautifully outlined in the first part of Aldous Huxley's *Literature and Science*,[48] a work written with an eye on the Snow-Leavis debate. It opens in fact with the question: "Snow or Leavis? The bland scientism of *The Two Cultures* or the violent and ill-mannered, the one-track moralistic literarism of the Richmond lecture?" The steps listed by Huxley aimed at reproducing in a manner as vivid as possible the uniqueness which is a chief characteristic of particular events experienced by man. With the literary man it is a perennial struggle to transcend the sphere of ordinary words, or rather to intersperse ordinary words with those that are rarely or hardly ever used. His is a steady search for new combinations of words, for arresting metaphors, and for unusual syntactic constructions. He revels in verbal sonorities, and does not disdain even recklessness with words if it can evoke some strange but real impressions. The literary man's manipulation of words and syntax puts him in a class very different from the authors of scientific papers where the standard is stark simplicity which best achieves the aim of scientific work: the demonstration of the common, invariable feature in physical processes.

While it may sound a truism, it cannot be repeated often enough that for the scientific investigation of nature the esthetic beauty in nature is largely irrelevant. The scientific analysis of a violet differs only in accidentals from the analysis of manure; scientific research is as unimpressed by the fragrance of the former as it is undisturbed by the stench of the latter. The physicist may at most be led to the concept of a bubble chamber from reflecting on the process which is described in Wordsworth's Inscription XII:

> Hast thou seen, flash incessant,
> Bubbles gliding under ice,
> Bodied forth and evanescent,
> No one knows by what device?

But the poet sees in the same process something which engulfs the whole man to the innermost core of his ability to feel. The poet can declare, and meaningfully so, about those vanishing bubbles that "such are thoughts." Moreover, the poet can repeat himself without becoming repetitious by adding a similar hue to his message. He does it by offering a parallel insight into the metaphorical significance of other phenomena of nature:

> Such are thoughts!—A wind-swept meadow
> Mimicking a troubled sea,
> Such is life; and death a shadow
> From the rock eternity![49]

The evocative power of the rock, to say nothing of that of a wind-swept meadow, has a sweep that knows no bounds. It reaches from the silence of solitary fields to the solitude of death and eternity. It is this range of evocations and impressions that cannot be accommodated in the categories of science. In science a rock is not even a rock, but only a case of wave functions and energy levels of the agglomerate of atoms and molecules. Ultimately a rock is for science a complex wave-number, however difficult to calculate exactly. Whatever the difference between a piece of gravel, of which there are millions on each country road, and the unique Sugar-Loaf towering over the Bay of Rio, in science their difference can easily approach the vanishing point.

No difference can indeed appear to be more vanishing than the one handled by numbers. Let the weight of that rock at Rio be expressed in tons and let it be a million or so tons, or 10^6 tons. The weight of a small pebble expressed in the same unit will be a millionth of a ton, or 10^{-6} tons. The scientific difference is now a mere dash in the exponent. It is the unique incisiveness of science that it can express immense differences in vanishingly small symbols such as a dash. But science never stated that the enormous difference in weight between the rock at Rio and the gravel on the road is equal to a small dash. Nor did science ever claim that weight in particular and measurable parameters in general make all the difference between those two pieces of rock made of the same elements. It was only some scientists dabbling in philosophy and some philosophers dabbling in science that tried to have our whole modern culture believe that this was the whole real difference.

No wonder that there has arisen under our very eyes a counter-culture, which is at times violently antiscientific. Its best spokesmen are poets, but none of the younger ones improved on what e. e. cummings of lower-case fame put down half a century ago in four lines of high pitch:

> While you and i have lips and voice which
> are for kissing and to sing with
> who cares if some one eyed son of a bitch
> invents an instrument to measure spring with?[50]

Such and similar instruments form the framework of cultures whose shocking features needed novelists for a realistic portrayal. *Brave New World* and *1984* are some of the best-known pieces in that literature out of which one can learn far more about threats to culture than from tons of science fiction. Cummings did not need to read anything to revolt. All he had to do was to keep his eyes open and assert his human sensitivity. This he did better than some latter-day poets of counter-culture. He was also better than most of them in his using four-letter words only sparingly. As was noted before, a poet, a man of letters, must at times resort to out-of-the-way words to let his readers realize a predicament of utmost seriousness. What is so serious is that after a hundred years since Huxley's lecture, the cultural crisis is far more severe than was suspected fifteen years ago when Lord Snow created a stir with his phrase, the two cultures.

About this seriousness Snow and other apostles of scientism could only give a few catchy phrases. When it came to serious reasoning, contradictions were not long to appear. In addressing in 1966 the Committee of the U.S. House of Representatives on Science and Astronautics, Snow claimed that men with wisdom though without scientific training were preferable to scientists without wisdom when it came to making vital policy decisions. Then he quickly added that this was not the real alternative.[51] Undoubtedly, it was possible to have scientifically trained men who were not short on wisdom. But the crucial point which Snow once more ignored was the source of that wisdom. Was it to be had by studying calculus or by reading the great classics of literature?

Revealingly enough, when Lord Snow was at his best, when he was most persuasive in *The Two Cultures,* he turned poet as he told his unforgettable stories. Poets have indeed a paramount and permanent role in culture. Only they can rise in telling indignation against those who try to measure quantitatively the beauty of spring, the value of ethics, and the weight of truth. Only poets have the verbal tools to ring a resounding alarm about the colossal robbery which is at the bottom of our cultural crisis and of its century-long split in two. The robbery is a sustained campaign to declare valueless any knowledge which is not quantitative, which is not expressed in terms of measurement. To say robbery is to make a very serious charge. It will be substantiated in the next lecture with a hint or two as to what makes knowledge full and therefore human in an age of science.

---

[1]References will be to the reprint in Thomas Huxley, *Science and Education* with an introduction by Charles Winick (New York: The Citadel Press, 1964), pp. 72-100.

[2]Justin McCarthy in 1899. See Cyril Bibby, *T. H. Huxley: Scientist, Humanist and Educator* (London: Watts, 1959), p. 67.

[3]Best known in this respect is the opening section of the *Pensées* on "l'esprit de géométrie et l'esprit de finesse," but emphasis on that difference forms a backbone of Pascal's reasoning throughout that work.

[4]See the conclusion of Chapter XV, "Géométrie" in his *Essai sur les élémens de philosophie,* in *Oeuvres philosophiques, historiques et littéraires de d'Alembert* (Paris: Bastien, 1805), vol. II, p. 326.

[5]The prophets were Turgot, Condorcet, Saint-Simon, Fourier, and Comte. For a full treatment of their social program, see F. E. Manuel, *The Prophets of Paris* (Cambridge, Mass.: Harvard University Press, 1962). For a discussion of their misuse of physics and astronomy, see my *The Relevance of Physics* (Chicago: University of Chicago Press, 1966), pp. 461-75.

[6]The principal miscomprehension of the Romantics about science was that they did not see the difference between science and the "scientific" claims of a mechanistic philosophy. Such miscomprehension gave rise among other things to Goethe's lifelong crusade against Newtonian optics. For details, see my article, "Goethe and the Physicists," reprinted above.

[7]Mentioned in the Presidential Address at the Oxford meeting of the British Association for the Advancement of Science; see *Report of the Ninety-Fourth Meeting, Oxford—1926* (London, 1926), p. 3. The scientists to be given honorary degrees were Brewster, Dalton, and Faraday.

[8]"A Liberal Education," p. 77.

[9]Ibid., p. 90.

[10]References will be to the reprint in Huxley, *Science and Education,* pp. 120-40.

[11]Ibid., p. 124.

[12]Ibid., p. 126.

[13]"The Function of Criticism at the Present Time" formed a part of Arnold's *Essays in Criticism,* published many times since 1865. For the quotation see the edition with an introduction by Clement A. Miles and notes by Leonard Smith (Oxford: Clarendon Press, 1918), p. 34. On Arnold's role in the educational reforms, see W. F. Connell, *The Educational Thought and Influence of Matthew Arnold* (London: Routledge & Kegan Paul Ltd., 1950). This work also gives valuable information on the position taken by the British Association concerning such reforms. See especially, p. 196.

[14]"Science and Culture," p. 127.

[15]Ibid., p. 135.

[16]Ibid., p. 125.

[17]Ibid., p. 136.

[18]Ibid.

[19]Ibid., p. 140.

[20]Ibid.

[21]"Literature and Science," *The Nineteenth Century* 12 (1882): 216-30.

[22]See J. F. Byrnes, *All in One Lifetime* (New York: Harper, 1958), p. 284.

[23]"Literature and Science," p. 219. The same phrase recurs in slight variations in the remaining part of Arnold's lecture.

[24]Such a formulation by Arnold of the chief point in Huxley's cultural program did full justice to the thrust of Huxley's aim.

[25]"Literature and Science," p. 224.

[26]Ibid., p. 225.

[27]Ibid., p. 227.

[28]"What Knowledge Is of Most Worth?" (1859), in *Education: Intellectual, Moral, and Physical* (New York: D. Appleton and Co., 1889), p. 40.

[29]See, for instance, Chapter 12, "The Barbarism of 'Specialisation'," in J.

Ortega y Gasset, *The Revolt of the Masses* (authorized translation from the Spanish; New York: W. W. Norton & Co., 1957). The Spanish original was first published in 1930.

[30]*The Two Cultures: And a Second Look* (Cambridge: University Press, 1969), p. 10. This is the paperback reprint of Snow's lecture expanded by an essay, "A Second Look," first published in 1964. The lecture was originally printed as *The Two Cultures and the Scientific Revolution* (Cambridge: University Press, 1959). The "Second Look" was Snow's reply to his critics. In general, scientists (A. C. Lovell, J. Cockcroft) endorsed his reasoning, and so did Bertrand Russell. Non-scientists (D. Riesmann, J. H. Plumb, N. Ayrton) found much fault with it. (See *Encounter,* August 1959, pp. 67-73.) M. Polanyi took Snow to task for his failure to see the limitations of scientific knowledge (*Encounter,* September 1959, pp. 61-62). One of the few scientists who strongly disagreed with Snow was F. Hoyle (*Of Men and Galaxies* [Seattle: University of Washington Press, 1964], p. 24), who noted that "creative spirit cannot be engendered by five-year plans." The validity of this remark was largely offset by Hoyle's "sociologism," by his identifying the source of creativity as the attention to the needs of society. Were Copernicus, Newton, and Einstein attending to social needs?

[31]*The Two Cultures: And a Second Look,* p. 10.

[32]A. V. Hill, *The Ethical Dilemma of Science and Other Writings* (New York: The Rockefeller Institute Press, 1960), p. 84.

[33]*The Two Cultures: And a Second Look,* p. 24.

[34]Ibid., pp. 23-24.

[35]Ibid., p. 24.

[36]Ibid., p. 28.

[37]Ibid., p. 31.

[38]Ibid., p. 36.

[39]Ibid., p. 44.

[40]Ibid., p. 45.

[41]In praising Snow's *Two Cultures* in 1959, Cockcroft still found nothing wrong with the fate of many industrial workers and scientists (and also of some sciences) in "Russia." He professed to know that at that time twice as much was spent there on pure research as in the United States. His figures could only be obtained by ignoring the obvious, namely, that in "Russia" private funds did not exist, whereas in the States they heavily supported pure research without being tabulated in the Government budget. Cockcroft did not seem to be aware of a statement of R. J. Oppenheimer which gained wide publicity: "In 1938 I met three physicists who had actually lived in Russia in the thirties. All were eminent scientists, Placzek, Weisskopf and Schein; and the first two have become close friends. What they reported seemed to me so solid, so unfanatical, so true, that it made a great impression; and it presented Russia, even when seen from their limited

experience, as a land of purge and terror, of ludicrously bad management and of long-suffering people." In *In the Matter of J. Robert Oppenheimer* (Washington, D.C.: U.S. Government Printing Office, 1954), p. 10.

[42]*The Two Cultures: And a Second Look,* p. 49.

[43]Henry David Thoreau, *Walden* (New York: Thomas Y. Crowell Company, 1961), p. 47.

[44]Herman Melville, *Moby Dick, or the Whale,* edited by Luther S. Mansfield and Howard P. Vincent (New York: Hendricks House, 1952), p. 185.

[45]B. Russell, *The Impact of Science on Society* (New York: Columbia University Press, 1951), p. 59.

[46]Originally printed in *The Spectator,* March 9, 1962, pp. 297-303. Reprinted in F. R. Leavis, *Two Cultures? The Significance of C. P. Snow* (New York: Pantheon Books, 1963). This volume also contains an essay by M. Yudkin, "Sir Charles Snow's Rede Lecture."

[47]L. Trilling, "Science, Literature and Culture: A Commentary on the Snow-Leavis Controversy," *Commentary* 33 (1962): 461-77.

[48]Aldous Huxley, *Literature and Science* (New York: Harper & Row, 1963). The next year saw the publication of *Science and Literature: Toward a Synthesis* by J. J. Cadden and P. R. Brostown (Boston: D. C. Heath, 1964).

[49]*The Poetical Works of William Wordsworth* with an Introduction by E. Dowden (London: Ward, Lock & Co., Ltd., n.d.), p. 435.

[50]From the poem, "voices to voices lip to lip" (1926) in E. E. Cummings, *Complete Poems 1913-1962* (New York: Harcourt, Brace, Jovanovitch, Inc., 1972), p. 264.

[51]See *Science* 151 (1966): 651.

7

Knowledge in an Age of Science

The heavy charge which brought the previous lecture to a close relates to a major robbery. Yet the very context of that charge may amount to at least a petty theft. The context—a hundred years of two cultures—could create the impression that the cultural split is about a century old. Such an impression would be very misleading. It would constitute a theft of a little over two thousand years. Two millennia may seem a small matter but when a charge is made about a major robbery it seems advisable not to become guilty of a minor theft even if it were merely academic.

There is an even more pressing reason to take a leap of two thousand years back into the past than merely the need to come clean academically about the whole story of two cultures. Where that story really starts, in the cave prison of post-Periclean Athens, there are also the origins of the topic of this lecture on knowledge in an age of science. This may seem almost absurd if one takes the view that science is a privilege of modern times. Yet while science is in a sense a truly modern achievement, thinking scientifically is a very old affair.

By thinking scientifically more is implied here than thinking rigorously. Thinking scientifically will now be taken in its more specific and genuine sense, namely, in the sense of thinking in

terms of matter and motion, in terms of purely spatial and temporal relations, all of which can easily be reduced to quantities.
Such a way of thinking is as old as Western thought whose origins
are usually assigned to the Ionian physicist-philosophers, Thales,
Anaximander, Anaximenes, and still later, Anaxagoras.

Anaxagoras was among other things the author of a book called
"The Mind,"[1] which made a tremendous impression on young
Socrates. In that book Anaxagoras described how changes follow
one another in the various parts of nature, in the sky, in the air, on
the earth, and even in the human body. To know, for instance, that
clouds follow hot days, rains follow clouds, floods follow rains, and
hot days will reduce floods, and so forth, was not only interesting
but also true. The sequence could only please one's mind and
seemed to reflect a higher mind that arranged and planned all
these changes in a nice, orderly fashion. Anaxagoras' physics
seemed to reveal even more recondite things. He claimed that the
big black stone that hit Aegospotami around 464 B.C. literally out
of the blue, was a chunk from the moon, or the sun, or some other
heavenly body. Clearly, this could appear far more pleasing to any
mind tuned to orderly sequences than the possibility that Jupiter or
some other capricious god hurled that stone down from Mount
Olympus. Socrates was very pleased, probably as much as a sophomore who for the first time understands the dynamics of atmospheric circulation and the eclipses of the moon and the sun. The
first is a purely temporal sequence, or so it may appear, and the
second is a purely spatial relationship of the sun and the earth. And
just like a sophomore, Socrates thought he now knew and understood everything that could be understood.

As years went by Socrates began to have doubts about his omni-
science based on Anaxagoras' physics. Had Socrates lacked depth
he would have never had second thoughts on the matter. But
Socrates was a person committed in an unusual measure to truth
and values. In his younger years he not only talked as most of his
friends did about democracy and freedom but also risked his life for
it. Twice he went to the battlefield and left service as a twice-
decorated war hero. When after Pericles' concern for liberty was
replaced in Athens by libertinism and the intellectual life began to
be dominated by the Sophists, Socrates started asking some apparently simple but very probing questions. While the Sophists
believed they had an answer for everything, Socrates professed to

know nothing. Needless to say, he did not become the darling of society. He ended up in the two-room prison carved into the northern slope of the Hill of the Muses across from the Acropolis.

His friends came and paid off the jailkeeper. He threw open the gates but Socrates showed no intention of escaping. Escaping, he argued, would create the false impression, a lie in short, that he felt guilty of the charge of corrupting the youth of Athens. The reply of his friends was that one's life was certainly worth a lie, especially if it harmed nobody. To this Socrates retorted that such was an expedient course only if life ended with death. There followed between Socrates and his friends the famous debate on the immortality of the soul which is movingly told by Plato in the *Phaedo*.[2]

The friends of Socrates argued against immortality on the ground that man was but a mere agglomerate of atoms. Man's knowledge about himself consisted in knowing the mechanism of the parts of his body, the purely quantitative correlation of pieces of matter. Since this was all the valid knowledge, any talk about the immortality of the soul and kindred topics was a mere waste of words. It was at this point that Socrates showed both his greatness and his great limitations. He wanted immortality, he wanted abiding sense of purpose, he wanted values, he wanted a lasting meaning of existence. He also saw that on the basis of Anaxagoras' physics, on the basis of purely quantitative considerations, he could not have meaning, purpose, values, let alone immortality. One could have these, he reasoned to his friends, only if a purely quantitative physics was wholly wrong. In its place had to come a physics which was about goals and purpose.[3] In the new physics proposed by Socrates, and fully developed by Aristotle, the chief question about things was not where they were and in what sequence they followed one another, but whether it was the best and truly purposeful for them to be in such and such correlation.

Such was the first dramatic scene in the long story of two cultures. From the very start the story was the story of what constitutes valid knowledge. It was Socrates' greatness to see that one form of knowledge, the quantitative, scientific, empiricist kind of knowledge gave no clue whatever about sense of purpose, meaning of existence, let alone about immortality. He saw this so clearly that he chose death in obedience to his conscience, an act that made him transcend human limitations. One limitation he could not transcend. He failed to perceive that he tried to remedy the short-

comings of one type of knowledge, the shortcomings of quantitative, empiricist reasoning, by another, equally limited type of
knowledge, the one based on reflection and introspection. The next
two millennia largely inherited this limitation of Socrates. The
inheritance was the almost exclusive concentration on questions
relating to purpose.

When one inquires about purpose one has to rely on introspection. No sensory experience, no external data can reveal purpose in
a strict sense. No wonder that the whole Aristotelian physics,
which was about purpose, was largely an unwarranted projection
of purpose into every phenomenon of nature. Purpose was pro-
jected into the fall of bodies, into the motion of stars, into the rise
of winds, even into the distinction of fresh and salt waters.[4] It is
therefore no wonder that the whole Aristotelian physics was an unbroken chain of erroneous propositions.[5] Again, it is no wonder
that reaction to it was as extreme as was Socrates' reaction to the
physics of the Ionian physicist-philosophers. Since the major error
of Aristotelian physics was the projection of purpose into everything, the reaction to it could only proceed by casting suspicion on
purpose in each and every form. This suspicion was heavily bol-
stered by the success of Newton's physics which was geared to deal
precisely with sensory or experimental evidence. Because of its
success it was tempting to reach the conclusion that only such
knowledge was valid which directly related to the sensory or quantitative. This conclusion, which was in the making within the
empiricist school from Ockham on, was spelled out by David Hume
with brute force at the end of his *Enquiry concerning Human
Understanding*. There Hume encouraged his readers to purge
libraries according to the precepts laid down in the *Enquiry*. His
readers, now persuaded Humeans, were to examine each volume,
especially volumes of theology and scholastic philosophy, and ask
about each two questions: "Does it contain any abstract reasoning
concerning quantity or number?" and "Does it contain any experimental reasoning concerning matter of fact and existence?" What
was to be done if the answer was in the negative? Hume's advice
was: "Then commit the book to the flames; for it can contain
nothing but sophistry and illusion."[6] In Hume's own words such
was the havoc to be played with libraries. The havoc was a confisca-
tion of books in a truly inquisitorial style and on a gigantic scale, a
wholesale robbery in short. It was to be committed by declaring a

major part of knowledge to be mere sophistry and illusion.

The wholesale burning of books as advocated by Hume may appear to have, on a cursory look at least, a mitigating circumstance. He seems to have been willing to spare books which in part at least were about quantitative reasoning and matters of fact. Whatever his willingness to spare such books from the holocaust he advocated, he certainly was in no mood to tolerate the kind of knowledge which was not about quantities and about matters of fact that did not lend themselves to quantitative treatment. Such was the only position permitted by Hume's radical empiricist starting point. If a philosopher's greatness is measured by his consistency then Hume was certainly one of the greatest. His empiricist position forbade him to accept any generalization as strictly valid from the viewpoint of reason. All generalizations were to be ascribed to instinct, a rather mysterious factor in itself and certainly in Hume's philosophy. No wonder that the realm of existence, as seen by reason, became for Hume a jumble of disconnected sense perceptions in space as well as in time.

Hume first perceived that this disconnectedness severed him from the world and from others. The First Book of the *Treatise concerning Human Nature*, the first version of the *Enquiry*, came to a close with the sad admission: "I am . . . affrighted and confounded with that forlorn solitude in which I am placed in my philosophy, and fancy myself some strange uncouth monster, who not being able to mingle and unite in society, has been expelled [from] all human commerce, and left utterly abandoned and disconsolate."[7] Somewhat later, he not only felt that his philosophy deprived him of unity with the rest of the world but also of the unity of his own self. There was no logical, rational guarantee on the basis of his empiricism that there was a David Hume, transcending the succession of the moments of his self-consciousness. In commenting on the merits of his theory of valid knowledge he wrote: "All my hopes vanish when I am to explain the principles that unite our successive perceptions in our thought or consciousness."[8] In the same Appendix to the *Treatise* he also admitted that when it came to the question of personal identity, "I find myself involved in such a labyrinth that, I must confess, I neither know how to correct my former opinions, nor how to render them consistent."[9] Empiricist philosophy could not be made consistent with the experience of personal identity, such was the major lesson of Hume's philosophy.

It stood for being trapped in a labyrinth of one's own making, but is not this always the result of a major robbery?

The time was now 1750 or so. As in the case of so many great philosophers Hume's originality lay in part in the incisiveness and consistency with which he articulated certain preferences that were already in the air. There were quite a few others who articulated the same preferences with less depth but with some catchy phrases and with at times stark crudeness. Since Hume merely drew the whole logic of Locke's *Essay concerning Human Understanding,* Voltaire was correct in stating that Locke had finally provided mankind with the true "history of the soul." In Voltaire's words Locke succeeded where all the others failed because he was "aided everywhere by the torch of physics."[10] Clearly, for Voltaire history of the soul in this case meant physics of the soul. The idea, "physics of the soul," was very expressive, but just as contradictory would it have been if Voltaire had described Newton's physics as the "soul of matter and motion." But when one becomes trapped in a labyrinth, contradictions are bound to rule. A generation or so later associationist psychology became the rage of the day and soon it became a custom with early British cultivators of psychology to call it "intellectual physics."[11] Did this imply that ordinary physics was not intellectual?

On the cruder side there was de la Mettrie with his *L'homme machine,* or *Man a Machine.*[12] In that book man was completely reduced to a mere machine in his individual as well as social aspects. In de la Mettrie's hands knowledge became a sheer compliance with the laws of physical nature. Or as Baron d'Holbach, an ally of de la Mettrie, put it in an extraordinary phrase: "All the errors of man are errors of physics."[13] De la Mettrie was also a chief starting point of a purely mechanistic biology which soon became the dominating trend in that science. Once biology and psychology began to be reduced to physics, nothing was more natural than to see the same trend assert itself in other fields as well. Indeed, Auguste Comte, the founder of sociology, saw the justification of this new science as a rigorous discipline in the fact that Newton's physics proved the existence of strict laws in nature.[14] While a static analysis of society could seem on a superficial look to be possible on the basis of physics, it was otherwise with the dynamic aspects of society or of its history. But here, too, the temptation was great to go physicalist, that is, to reduce one's field of inquiry to the methods of physics. Already Voltaire tried to write a history of France as exact as the physics of Newton. A century or so later Walter Bagehot, a sociologist and economist of great renown in his time, came out with a book, entitled *Physics and Politics*.[15] There Bagehot argued that the chief force in history and politics was unconscious imitation on the part of man. Whether this is true is another matter. It is certainly true, however, that consciousness is not the subject matter of physics and that therefore the setting of conscious aims, or a conscious pursuit of ideals and goals, could not be a factor to be considered in the science of politics and history.

This lure of physics, of its exactness, of its apparently pure empiricism, is known as physicalism.[16] Whether one likes it or not, one has to admit that physicalism has become a dominating feature of modern intellectual history. In 1909 Henry Adams had already diagnosed this situation in an essay whose title should make anyone sit up who is still not completely brainwashed by physicalism, this modern credo of academia. "The Rule of Phase Applied to History," so reads the title of an essay which is an odd mixture of terms of physical chemistry and of historiography. The essay might very well be forgotten except for one phrase: "The future of thought and therefore of history lies in the hands of physicists, and therefore the future historian must seek his education in the world of mathematical physics."[17]

One of the latest things one can learn from mathematical physics is the astonishing range of possibilities of data processing by computers. No wonder that such opportunities turned a head or two. Half a century after Henry Adams, a prominent student of Civil War history proposed that the whole voting record of the Congress between 1825 and 1859 be computerized. Only then, he claimed, shall we know what caused the Civil War.[18] One wonders if Abraham Lincoln is not turning in his grave on seeing taxpayers' money being channelled into such and similar projects. Should one indeed believe that the role of England in World War II will not become clear until computers have tabulated the relative frequency of adjectives in the speeches of Winston Churchill? And what about those trite adjectives which when used by Churchill could electrify not only his audience but the whole Free World? One of those uses took place in Ottawa where he dismissed the Nazi threat to wring the neck of the British chicken with the words: "Some chicken, some neck." Computers will not be a match for trite words that proved themselves so powerful in defying entire armies. Let us therefore hope that the study of history will not be heavily dominated at least in the Free World by these physicalist historians. And even if this were to happen, it would only provide one more illustration of the old saying: History does not repeat itself, only some historians repeat one another.

Where the repetition of physicalist slogans is almost complete today is in the study of psychology and sociology. The slogans are not called physicalist. They usually go under the label of behaviorism. Yet the founders of behaviorism had been very clear that what they aimed at was pure physicalism: "The essence of behaviorism," K. S. Lashley wrote in 1923, "is the belief that the study of man will reveal nothing except what is adequately describable in the concepts of mechanics and chemistry,"[19] that is, if a small explanation is allowed, in terms of exact physical science. A year later, J. B. Watson, another founder of behaviorism, disclosed somewhat unwittingly how such a program would put things on a runaway course: "The interest of the behaviorist is more than the interest of the spectator: he wants to control man's reactions, as physical scientists want to control and manipulate other natural phenomena."[20] This is the program of control for control's sake, and it can only trigger a runaway process. Its end is the brave new world described by Aldous Huxley, or a world which is beyond freedom and dignity, as reads the title of a not-so-old book of B. F. Skinner, a chief spokesman of present-day behaviorism.

If there is a runaway course it is from heaven to hell. The promise of behaviorism was a heaven on earth, the heaven of science. Forty years ago, Professor E. G. Boring of Harvard University wrote this in the Preface of his once famous book, *The Physical Dimensions of Consciousness*: "The ideal would be ultimately to get away from conscious dimensions to the happy monism of the scientific heaven."[21] By monism he meant the exclusive validity of quantitative knowledge, of quantitative laws, physicalism in short. In that heaven one could perhaps still have consciousness but one is not supposed to say anything about it. A strange heaven indeed where you cannot even say that you are there. But spokesmen of behaviorism often speak as if the principle of contradiction did not exist for them. Others who are not behaviorists by profession often do not realize that they are caught in a contradiction essential to behaviorism when they try to straighten out society. The police certainly ought to behave like angels. But should policemen behave like angels in a society where intellectuals only smile at belief in angels, that is, at belief in unconditionally valid moral principles? If behaviorism is true, what forbids that the police should adopt a behavioral pattern characteristic of thieves and robbers? Clearly, it is too late to pay more than lip service to unconditionally valid moral principles when the crime rate is rising so rapidly as to make law enforcement impossible.

The behavioristic collapse of society cannot be rigorously objected to on the basis of behaviorism. Al Capone is known to have boasted that all his actions were but the full utilization of the opportunities provided by what he called "the system."[22] Had Al Capone been a philosopher, he would have said that he meant a society that borrowed its norms from behaviorism. Had he lived longer he would have found no small satisfaction in the behavioristic course taken by Supreme Courts everywhere in the Free World. But here it is not so much ethics as valid knowledge which is in question. To cut through the maze of philosophical abstractions and distinctions it is best to take a look at the physiognomy of a truly behavioristic piece of information. But since behaviorism is an effort to give only the scientifically factual by eliminating everything allegedly metaphysical and mythological, let that piece of information be first seen in its old-fashioned form:

> The Lord is my shepherd, I shall not want.
> In green pastures he gives me repose.
> He leads me to waters where I may rest.
> He guides me in right paths for his name's sake.
> Though I should walk in a dark valley,
> I will fear no evil.
> For you are at my side with your rod and your staff
> That give me courage.

In its behavioristic version the same is reduced to the following:

> The Lord is my external-internal integrative mechanism.
> I shall not be deprived of gratifications
> For my viscerogenic hungers or my need dispositions.
> He motivates to orient myself
> Toward a non-social object with effective significance.

> He positions me in a non-decisional situation,
> He maximizes my adjustment.

The first reaction to all this may be a burst of laughter and this behavioristic variation of Psalm 23 was originally meant to be a little farce when composed in the early 1960s.[23] But it is well to remember that all good comedies have a serious message as do all good cartoons. The message is the reassertion of the normal by setting it off against the grotesque. Human nature is probably too stable to assimilate on a large scale the bizarre precepts of behaviorism. Yet the danger should not be underestimated. What may be a joke today can be a tragedy tomorrow. On the face of it, it is only a college prank when an engineering student submits an analysis of Hamlet in terms of physical science. It happened in an exam in English literature at M.I.T., and the essence of the analysis was that Hamlet had too much feedback in his circuits and therefore he worked at a low efficiency. Thus he caused the death of six persons, although all his problems could have been solved had he killed his own mother. The ratio of one to six is 16.66 percent, worse than the efficiency of an ordinary combustion engine.

The story, a true story,[24] is humorous enough, but on second thought it can also send a chill down one's spine. It seems inconceivable that a generation, let alone a century ago, a young university student would have thought of taking a similar approach to one of the greatest pieces of literature. There were pranks even in those days but somehow the general academic atmosphere did not invite the kind of prank which occurred to our engineering student. Even a hundred years ago it was well known that it could be said of man that he was among other things an internal combustion engine. But somehow it could not pass even for a joke to suggest that Hamlet was one such engine and that this was all the valid knowledge one could have about him. This is not to draw attention from the fact that the greater is the concentration of the mind, the greater is the consumption of calories in the brain. But calories, however carefully calculated, are not for the purpose of conveying the depth of thought and the gripping seriousness by which tragedies embody the struggle for purpose and the quest for the ultimate meaning of existence.

Of such tragedies, of the great human drama which is acted out in each human psyche, a behavioristic psychology can know nothing as long as it remains consistent with its physicalist presuppositions as to what constitutes valid knowledge. Such a psychology does not even have the right to be called a study of the psyche as it recognizes only external behavior. The only thing such a psychology can teach is the manipulation of human beings by taking advantage of their various weaknesses. Such a psychology cannot give norms and ideals except such as are popular at a particular place and at a particular time. What such a psychology decries with a crusading spirit is the reference to the absolute and to the unconditionally valid. Such a psychology does not want to know whether a war is just, but only whether it is popular or not. Such a psychology is ready to relativize even the crimes of genocide. It is willing to write off with calm scientific objectivity the five million peasants killed in Stalinist Russia, the six million Jews exterminated in Nazi Germany, and the still uncounted millions of people purged in the People's Republic of China, to mention only the larger crimes committed in this century of genocides. They can hardly appear crimes to a psychology which claims that the notion of crime is merely the function of a particular social conditioning.

There are a good number of reasons for rising in indignation against such a "sophisticated" handling of crime in general and of genocide in particular. One of these reasons is the claim of behaviorism that its theory of valid knowledge is the logical derivative of science, of exact science. Such a claim is almost as old as science itself. No sooner had exact science reached its full maturity with Newton than a purely empiricist interpretation of existence gained immense popularity. Popularity it was also in the sense that all its supporters were people whose familiarity with exact science hardly ever transcended the popular level. On that level everything was a machine or part of a machine and that was all that could be known. The kind of knowledge which made real exact science, such as that of Newton's *Principia,* was on a much higher level. It is, however, well to remembr that those chief advocates of mechanistic and empiricist philosophy, Locke, Hume, Voltaire, and others, had not enough scientific training to read Newton's *Principia,* but only Locke had the honesty to admit this.[25] The others tried to parade as experts in Newtonian science. Due to their ignorance of that great and hard book they could blissfully overlook the fact that Newton's creativity in science implied a theory of knowledge which far transcended the shallows of empiricism or mechanism.[26] What they could not ignore was that Leonhard Euler, the greatest exact scientist in post-Newtonian times, was fully aware of the wanton exploitation of exact sciences on behalf of empiricism and mechanism. To him it was wholly incomprehensible and unjustified that the usefulness of a particular machine should be credited to the machine and not to the creative mind of the scientist who designed it. In this age fooled by the myth of thinking machines his words should have an uncannily modern ring: "Clever as the constitution of a machine may be, the praises accorded to it must devolve on the engineer who designed it. The machine itself has no concern whatever for praise; it is the engineer who is responsible for the faults of a clumsy and badly constructed machine; the machine itself is wholly innocent; bodies as such are responsible for nothing; no punishment, no reward is relevant to them; all changes and motions produced in machines are but the necessary consequences of their construction."[27]

The nineteenth century shows a similar contrast between men of science and some philosophers who kept invoking exact science on behalf of their mechanistic empiricism and materialism. Mention has already been made of Comte and Bagehot, but there were a number of others. There was, for instance, Marx, who in the Preface to *Das Kapital* declared that the laws of social and economic evolution were as rigid and inevitable as the laws of mechanics.[28] If such was the case why was it still necessary to work, and in a most purposeful way, for the advent of the dictatorship of the proletariat all over the world? Knowledge, according to the Hegelian left, is a mere surface phenomenon of social dynamics, which again, if dialectical materialism is true, is the inevitable manifestation of the transformation of matter. In that case, why was it again necessary on the part of Engels to engage in such a purposeful and willful activity as the writing of his book, *The Dialectics of Nature,* or his theory of scientific knowledge? In that book all scientists fared well except the great ones. Engels called Newton an inductive ass, and used equally derogatory words about Lord Kelvin, Tait, Helmholtz, and Clausius, a list that reads like the blue-ribbon panel of nineteenth-century physics.[29] Engels' hero and master in exact science was none other than Hegel, possibly the least exact among all who wrote extensively on the theory of knowledge.

Engels' knowledge of physics, especially of nineteenth-century physics, was very inexact indeed. Of its many gaps two illustrations should suffice. Engels did not mention Cauchy, who in 1821 succeeded in giving an exact formulation of the theory of the limit, the cornerstone of calculus and the very basis of exact, mathematical physics. Prior to 1821 great physicists, Lagrange for example, kept telling their students that they must take calculus on faith and wait until the rigorous proof of the limit would come. It is not likely that Engels would have found Cauchy too much to his liking had he known of Cauchy's epoch-making book containing the rigorous proof of the theory of limit. In its Preface Cauchy took pains to emphasize that calculus was not everything and that it would be a grave error to think that all valid proofs should be based on integral and differential equations. Such was a daring statement especially in the France of those times where graduates of the École Polytechnique occupied in large numbers high civil service posts and were busy in introducing the spirit of infinitesimals into politics. But as Cauchy wrote: "Nobody has up to now tried to prove by calculus the existence of Louis XIV; yet all in their right mind agree that his existence is as certain as Pythagoras' theorem. . . . What I have said of a historical event, can be applied equally well to a great number of questions, in religion, in ethics, in politics. Therefore, let us remain convinced that there are truths other than those of geometry, and realities other than those of sensible objects." His concluding advice was: "Let us therefore cultivate with fervor the mathematical sciences, without wishing to extend them beyond their range; and let us not imagine that one could attack the problems of history with mathematical formulas, or that one could sanction the principles of morality by theorems of algebra and calculus."[30]

Maxwell, another great name of nineteenth-century mathematical physics, was referred to repeatedly by Engels, who claimed that Maxwell's electromagnetic theory firmly established the existence of the ether and consequently the truth of materialism. Engels was not competent to realize that Maxwell's equations could be considered invariant only if one adopted relativity, a theory which still looks disturbingly non-materialistic to Party philosophers. Moreover, while Maxwell believed in the existence of the ether, he also believed in God. In particular, he was a physicist, and one of the greatest, who reflected at length on what was implied in the kind of cogitation done in exact science. It was no accident that he lampooned Hegelian metaphysics as well as Tyndall's scientific materialism, and even some misguided Christian men of science who looked for evidence about the soul in thermodynamics. It was such reflections that prompted him to make that statement which should be carved into each and every desk in each and every laboratory and science classroom: "One of the severest tests of a scientific mind is to discern the limits of the legitimate application of scientific methods."[31]

About the severity and seriousness of this test nothing is perceived by those who glibly claim that quantitative, empiricist knowledge is the only valid knowledge especially in an age of science such as ours. Curiously, it never dawns on them that it is this misguided, reductionist theory of knowledge which is a principal source of our present-day cultural distress. Nor did the Lashleys, the Watsons, the Borings, and the Skinners ever so much as hint that the most outstanding exact scientists of our own times raised their voices repeatedly against the unwarranted projection of the method of physical science into each and every area of inquiry. A curious silence about a striking contrast! On the one side there is the glib behaviorist with his electronic gadgets and maze-running mice, thinking that he has the infallible clue to that greatest mystery in nature, the nature of man. On the other side you see an Einstein, one of the greatest physicists of all times, who was not ashamed to confess his grave limitations not only as a man but also as a theoretical physicist. He was already the world famous father of special and general relativity when he wrote to Ehrenfest: "How miserably do we theoretical physicists stand before nature, and before our own students!"[32] In a letter written in 1952 he spoke of his theory of knowledge as implied in his scientific creativity. The theory was such that he, the professed agnostic, felt the need to add the quick aside: "Please, do not think that I have fallen in the hands of priests."[33] Well, he did not fall in the hands of priests, but he was too great a thinker not to perceive that his ideas on knowledge had an uncanny resemblance to an epistemology and metaphysics, which for the last three centuries has been advocated only by priests who knew what kind of epistemology was compatible with their belief in a personal Creator.

Planck, the creator of quantum theory, the other great branch of modern physics, is an equally eloquent witness against a physicalist theory of knowledge. He opposed positivism and empiricism whenever he could. His clashes with Mach were particularly sharp. In those clashes Planck advocated nothing less than the scientist's need of faith. To this Mach retorted that he had no taste for joining a church.[34] Of course, the church in question was neither Catholic nor Lutheran. The church, to which Mach referred, related to the fact that for Planck the rationality of nature was not, as Mach wanted it, a mere economy of thought, but a genuine assertion of a world which was rationally ordered and existed as such independently of the thinking mind. This rationality, so Planck insisted, could only be recognized through an act of the mind which was akin to an act of faith.[35] As in an act of faith the mind had to make a leap beyond the immediately given data of evidence. It should now be clear why Mach felt that accepting such a theory of knowledge was in a sense like joining a church. It is now also clear that precisely because of his refusal to join the church of "believing" physicists, Mach was unable to believe in the greatness of Planck and Einstein. Mach, the scientist, looked backward not forward. It was his opponent, Planck, who did creative work in science precisely because he had a much broader notion of knowledge and rationality than the notion permitted on the basis of Mach's empiricism.

Next to Planck and Einstein, the physicist to be mentioned most naturally is Niels Bohr, the one who unlocked the mystery of the atom. In part because of his speculations on atomic theory Bohr advocated a theory of knowledge known as complementarity.[36] The matter and wave aspects inherent in atomic theory confirmed his earlier conviction that reality revealed to the inquiring mind different aspects which could not be reduced to one another. A full knowledge could only be acquired when the various, mutually complementary aspects of reality had been carefully established and correlated. Like Planck and Einstein, Bohr was a resolute antireductionist.[37]

The twentieth-century scientific scene is full of prominent physicists who wrote on scientific method and knowledge in the general context of culture. Whatever their differences they all agree on one major point: They are not empiricists. Even Bridgman, a vocal supporter of operationism, was willing to contradict it when giving his definition of scientific method. There was no such thing, he declared. Scientific method was merely "doing one's damndest with one's mind."[38] He certainly wanted no part of empiricist cubbyholes. For Eddington this very same broader view of knowledge consisted in the fact that, as he put it, man's mind was not merely a measuring device. Reality as reflected in human knowledge displayed two irreducible aspects, the metric, or quantitative, and the non-metric, or qualitative. He went even further. He saw the ultimate justification of this broader view of knowledge in introspection, the very anathema to every empiricist and behaviorist. "It is by looking into our own nature that we first discover the failure of the physical universe to be co-extensive with our experience of reality."[39] In other words, man's experience and perception, in short, his culture, embraced a realm far wider than the field of quantities.

The introspective look as a valid form of knowledge was perhaps most strikingly articulated by Arthur Holly Compton of Compton-effect fame. What he said should seem particularly revealing in view of some remarks made earlier in this lecture about purpose. There it was claimed, and with an eye on Socrates, that man knew of purpose precisely because of his introspective ability. This Socratic claim may be suspect, precisely because of Socrates' destructive influence on physical science. But his basic insight about the primacy of our being conscious of acting freely and on purpose was fully vindicated by Compton. This is not to suggest that Compton had Socrates in mind in his Terry-lecture given at Yale in 1931. But what he said there on the problem of free will with respect to the laws of physics amounted to an unwitting justification of Socrates. In noting the steady erosion of confidence in man's inner sense of freedom, Compton remarked: "It seems unfortunate that some modern philosopher has not forcibly called attention to the fact that one's ability to move his hand at will is much more directly and certainly known than are even the well-tested laws of Newton, and that if these laws deny one's ability to move his hand at will the preferable conclusion is that Newton's laws require modification."[40] It is indeed unfortunate that it fell to a physicist to call attention to this non-physical and elementary truth. This truth is wholly missed by behaviorists of all kinds who spend so much time and energy on arguing against the freedom of will and the reality of purpose in the name of science. They certainly do not seem to expend all that effort either out of chance, or out of necessity, or out of both. These misguided efforts are consistent only to the extent that purpose and freedom are two sides of the same coin. You cannot have one without the other, nor can you deny only one and not the other as well. No purposive action properly so called can be had without free will and no free act can be carried out without some conscious purpose. About the purposive, wilful efforts to discredit purpose it is always refreshing to recall the words of Whitehead: "Scientists animated by the purpose of proving that they [their actions] are purposeless constitute an interesting subject for study."[41]

Unfortunately, that study when done in its broader ramifications is not only interesting but also reveals some tragically serious aspects of our culture. One of these great tragedies was World War II. It was certainly included in that appraisal which W. Heitler, one of the chief architects of quantum mechanics, offered in 1949: "In the decline of ethical standards which the history of the past fifteen years exhibited, it is not difficult to trace the influence of mechanistic and deterministic concepts which have unconsciously but deeply crept into human minds."[42] Politics played, of course, a part in bringing World War II about. But were not those politics, rank aggressiveness on one side, selfish indifference on the other, and dialectical calculation on still another, based ultimately on sheer opportunism in which there is no room for ideals, for responsibilities, but only for soulless maneuvering? The fifteen years following Heitler's dictum were not marred by a cultural cataclysm comparable to World War II. But the erosion of moral values went on, and the mechanistic factors were again clearly at play. In 1963 Erich Fromm summed up the situation in the following words at the meeting of the American Orthopsychiatric Association in San Francisco: "Man sits in front of a bad television program and does not know that he is bored . . . he joins the rat race of commerce, where personal worth is measured in terms of market values, and is not aware of his anxiety. . . . Theologians and philosophers have been saying for a century that God is dead, but what we confront now is the possibility that man is dead, transformed into a thing, a producer, a consumer, an idolater of things."[43]

It should seem obvious from the words used by Fromm, such as television, market, production, and the like, that the kind of idolatry he had in mind would be inconceivable without modern science and technology. Yet neither science, nor technology are responsible for this idolatry and for its sinistrous effects. Long before there was even a trace of science man worshipped things whether they were the products of his hands or not. The kind of idolatry made possible by science merely shows that unless man is possessed of moral values and moral strength science may make him oblivious either to those values or to that strength. It was not science that claimed that science can produce heaven on earth. Yet some philosophers and publicists enjoyed making such claims. They were the ones who created the belief that scientific knowledge was all that could be had in the way of valid knowledge. In addition to these philosophers and publicists there were some psychologists and biologists who tried to create the impression that they alone were talking in the name of exact science.

Curiously, the ones really entitled to speak in its name, the truly great theoretical physicsts, took a very different view. It was the Nobel Prize winner Polykarp Kusch who told Pulitzer Prize jurors in 1961 that "science cannot do a very large number of things." This sounded iconoclastic enough but a minor matter in comparison to what he added: "To assume that science may find a technical solution to all problems is the road to disaster." On such a premise it was quite logical to call for a cultural crusade. "I am quite certain that the mass of men believe that the better world of tomorrow will come through science. I think that such belief ought to be publicly combated."[44]

Whether it will be combated, and by responsible people, is still to be seen. To call attention to the limitations of science is a very touchy matter. It can easily expose one to the charge of obscurantism. It should therefore seem to be of no small help that those who did most in ushering in both the theoretical and the technological side of this age of science did not give their vote to the proposition that knowledge of quantities was alone valid knowledge. On the technological side few did as much as Vannevar Bush, the father of modern computers. Yet he was the one who wrote: "Much is spoken today about the power of science and rightly. It is awesome. But little is said about the inherent limitations of science and both sides of the coin need equal scrutiny."[45]

This equal scrutiny of both sides should make it clear that the central issue in our culture is the question of knowledge in an age of science. This question makes sense only if one is ready to recognize that knowledge is not a one-track proposition. In other words what ought to be recognized is that no specific form of knowledge can do full justice to the entire gamut of human experience about existence. Without this recognition a fruitless and sterile debate will go on between the humanists and the scientists. A recent instance of this is provided in the dozen or so papers that tried to shed light on the changing relationship between science and the general public.[46] The change in question is obvious. It consists in a growing dissatisfaction with science for its failure to give more than science can give. The change is a reaction to scientism, to the belief that science will redeem man and society of any and all problems. Yet as in the case of any other addiction, the core of scientism is sought in still more scientism. It is wholly ignored that the credo of scientism has never been supported by the truly great figures in exact science. It is completely missed that the faith of scientism is a most inexact rendering of what the knowledge provided by exact science and its method can do for man.

Even in its own field exact science cannot play the prophetic, redeeming role. Its prognostications as to what is going to be achieved in its own field are notoriously inexact.[47] As to its own problems, it never solves any major one without stumbling on a new and apparently even more intractable problem. Exact science certainly cannot solve problems which need attention and insight of another kind than what is involved in the handling of quantitative relationships. Only by ignoring this elementary truth can one make dramatic, but actually rather hollow, appeals to science to provide the answer to man's groping with the riddle of existence. This groping is not scientific but philosophical and religious. It transcends the realm of physics simply because it has to do with metaphysics. The answer to that groping is therefore to be had either by the tools of metaphysics and religion or it will not be had at all. Science may one day furnish a most exact timetable for the presence of the human species on the earth, but the amount of time thus specified will not be an answer to existential gropings. Science may one day describe with great accuracy the energy levels of all electrons in the brain while the individual is in the grip of existential questions, but the complicated mathematical formulas and the consciousness of having existential questions will remain bafflingly different in character.

Faced with that enormous difference of knowledge about the same phenomenon, one can do only two things. One can take it seriously or not. In the latter case one has to declare that it is meaningless to ask the question why such and such energy levels of billions of electrons evidence themselves as consciousness. Such a declaration does not deny that there is consciousness, but it certainly denies that it is meaningful to press on with the question why conscious experience is so enormously different from the equations trying to express it quantitatively. It should be easy to see that such a position affects the evaluation of each and every conscious experience including each and every bit of scientific knowledge actually known by the individual. The position, which in its most recent form is known as the psycho-physical identity theory,[48] declares about each such experience that our knowledge of its bafflingly non-physical character is a knowledge about which no further questions can be asked. This is, of course, equivalent to the declaration that such a knowledge is a largely useless knowledge.[49] But if such is the case, then even the question about the meaning of existence becomes meaningless because such a question cannot be logically raised if the specific difference between conscious experience and its scientific description is not a meaningful knowledge.

Humanists who call upon science to perform a duty it cannot perform are unaware of the fact that their appeal to science is based on the tacit acceptance of the tenets of the psycho-physical identity theory about valid knowledge.[50] These tenets constitute a wholesale robbery on the level of knowledge, because they are tantamount to the declaration that the deepest aspect of each bit of conscious knowledge is meaningless. To realize that these tenets are beset with self-contradiction is certainly a useful knowledge, but is not perhaps the most vital information in this connection. What one should be aware of above all is that investing science with a prophetic and messianic role has not been the doing of science. Exact science, or rather its best cultivators, have never claimed that role. Exact physical science came into its own when during the seventeenth century it eliminated from its ken questions about existence, meaning, purpose, and the like. No wonder that sensitive physicists instinctively reject appeals from shortsighted humanists to do science in a so-called meaningful, or prophetic way.[51] The cultivation of that meaningfulness is the business of the philosophy of being, or metaphysics, and of religion, if one wants to go even farther. This is not to suggest that science is not full of philosophical presuppositions. But philosophy as such is not a direct part of the scientific strategy of exploring what can be known quantitatively about nature and existence.

Still, the scientific strategy, however limited in its scope, is a most integral part of our humanity. The restriction of the word humane to the study of literature is in fact the sad fruit of the myopia which many a Renaissance man had toward science. True, science was not fully born yet when Ficino mesmerized the West with the myths of Plato, and when Erasmus did the same with his collection of catchy phrases from Greek and Latin authors. But even then enough could be foreseen about the future greatness of science to give it serious attention and respect. It could also be known already at that time that the scientific quest, however distrustful of philosophical preoccupations, engaged the energies of man in an eminent degree. The scientific quest is therefore as humanistic as is concern for poetry, music, letters, and politics. Like these it can engage only a part of man, it can satisfy only one aspect of man's ability to know.

This is not so only since the rise of science. Man has always been a philosopher as well as a scientist. This is why he survived. In that survival he needed mental vision as well as physical energies. The latter he first acquired, and perhaps over half a million years, by gathering fruit. But he had to do it selectively, that is, scientifically. It is an old truth that gathering fruit indiscriminately ends in death. In that long struggle for survival man made instruments and he painted as well. It was only very recently that man found the art of writing. The invention of science followed the invention of letters within two thousand years, a mere second on the evolutionary scale. What this close sequence suggests is that man has always been a man of letters as well as a man of science. He always needed to be both. The recognition of this is the kind of knowledge which is most needed in an age of science. It is needed on the part of scientists, as well as on the part of theologians, poets, and politicians. Once these different and equally needed professions become aware of the limitations of their own trade, the first step has been made toward the unity of knowledge in its diversity, which is the chief issue about culture in an age of science.

---

[1] For an informative reconstruction of Anaxagoras' physics and cosmology, see D. E. Gershenson and D. A. Greenberg, *Anaxagoras and the Birth of Physics* (New York: Blaisdell Publishing Company, 1964).

[2] That the *Phaedo*, certainly remembered by posterity as a major document on the immortality of the soul, played a decisive role in the history of

physics, is a fact almost invariably overlooked by historians of science. It was under the influence of Socrates that Plato and especially Aristotle developed the notion of organism as the "true" explanation of the physical world, an explanation which dominated scientific thought for about 2000 years. See on this Chapter I, "The World as an Organism," in my *The Relevance of Physics* (Chicago: University of Chicago Press, 1966).

[3]See sections XLV-LXII in *Phaedo*.

[4]For further details, see *The Relevance of Physics*, pp. 18-32.

[5]In the phrase of E. T. Whittaker, Aristotle's physics was "worthless and misleading from beginning to end." *From Euclid to Eddington: A Study of Conceptions of the External World* (Cambridge: Cambridge University Press, 1949), p. 46.

[6]See, for instance, the Gateway edition (Chicago: Henry Regnery Company, 1956) with an introduction by Russell Kirk, p. 173.

[7]*The Philosophical Works of David Hume* (Edinburgh: printed for Adam Black and William Tait, 1826), vol. I, p. 335.

[8]Ibid., vol. II, p. 559.

[9]Ibid., p. 556.

[10]"Lettres sur les Anglais ou Lettres philosophiques" (1733), in *Oeuvres complètes de Voltaire* (Paris: Delangle Frères, 1826-34), vol. XXXV, p. 95.

[11]See, for instance, Thomas Brown's *Lectures on the Philosophy of the Human Mind* (Philadelphia, 1824), vol. I, p. 120. On the continent Charles Bonnet graced his *Essay analytique sur les facultés de l'âme* with such chapters as "The Physics of Reminiscence," "The Physics of Imagination and Memory," and "The Physics of the Composition of a Discourse." See *Oeuvres d'histoire naturelle et de philosophie de Charles Bonnet* (Neuchâtel, 1872), vol. XIII.

[12]*Man a Machine*, French text and English translation with notes (La Salle, Ill.: Open Court, 1912).

[13]*Système de la nature ou des lois du monde physique et du monde moral* (new ed.; London, 1775), p. 19.

[14]This was a leading idea of the *Cours de philosophie positive* as well as of the *Système de politique positive*. See *The Positive Philosophy of Auguste Comte*, translated by H. Martineau (London, 1875), vol. I, p. 118.

[15]*Physics and Politics; or, Thoughts on the Application of "Natural Selection" and "Inheritance" to Political Society* (New York: D. Appleton, 1873). The subtitle of Bagehot's work is a telling reflection on Darwin's own effort to cast evolutionary theory into a purely physicalist mold.

[16]Physicalism obtained a programmatic formulation in the Vienna Circle, especially through the efforts of R. Carnap, as shown in *The Unity of Science* (London: Kegan Paul, 1934), the English translation by M. Black of Carnap's "Die physikalische Sprache als Universalsprache" (1931).

[17]Henry Adams, *The Degradation of the Democratic Dogma* with an intro-

duction by Brooks Adams (New York: Macmillan Company, 1920), p. 283.

[18]L. Benson in a lecture "Quantification and History" given at Princeton University on May 11, 1967.

[19]"The Behavioristic Interpretation of Consciousness," *Psychological Review* 30 (1923), p. 244.

[20]*Behaviorism* (New York: The People's Institute Publishing Company Inc., 1925), p. 11.

[21]New York: Dover Publications Inc., 1963, p. xiii (Preface to the 1933 edition).

[22]"The American system of ours, call it Americanism, call it capitalism, call it what you like, gives each and every one of us a great opportunity if we only seize it with both hands and make the most of it. My rackets are run on strictly American lines, and they are going to stay that way." Quoted in *National Review,* Dec. 20, 1974, p. 1444. While the Founding Fathers of the United States were far from intending to provide for such opportunism, Al Capone correctly sensed that when commitment to unconditionally valid ethical values seriously weakens, the stallion of liberty becomes the Trojan horse of libertinism.

[23]The text is taken from *Time,* August 9, 1963, p. 56, col. 3, where A. Simpson, President of Vassar College, and formerly professor at Washington University in St. Louis, is given as the author. He obviously meant to make his students aware of the self-defeating consequences of behaviorism.

[24]Reported by Prof. E. E. Morison in M. Greenberger (ed.), *Computers and the World of the Future* (Cambridge, Mass.: M.I.T. Press, 1964), p. 17.

[25]Locke consulted Huygens, who assured him that "all the mathematical Propositions in Sir Isaac's *Principia* were true." The source of this information is one of Newton's acquaintances, Jean-Théophile Desaguliers. See his *Course of Experimental Philosophy* (3rd ed.; London: A. Millar, 1763), p. viii.

[26]See Chapter 6, "The Instinctive Middle," of my *The Road of Science and the Ways to God* (Chicago: University of Chicago Press, 1978).

[27]*Lettres à une princesse d'Allemagne sur divers sujets de physique et de philosophie* (Lettre LXXXV, le 16 Décembre, 1760; Mietau et Leipsic: chez Steidel et Compagnie, 1770), vol. II, p. 23.

[28]*Capital: A Critical Analysis of Capitalist Production,* translated from the third German edition by S. Moore and E. Aveling (New York, 1889), pp. xxx-xxxi (author's preface to the second edition).

[29]*Dialectics of Nature,* translated by C. Dutt (New York: International Publishers, 1940), p. 155.

[30]Augustin-Louis Cauchy, *Cours d'analyse de l'Ecole Royale Polytechnique, Ire Partie, Analyse algébrique* (Paris: de l'Imprimerie Royale, 1821), pp. vi-vii.

[31]"Paradoxical Philosophy," in *The Scientific Papers of James Clerk Maxwell,* edited by W. D. Niven (Cambridge, 1890), vol. II, p. 759.

[32]The date of the letter is March 15, 1922. Quoted with the kind permission of the Einstein-estate.

[33]*Lettres à Maurice Solovine,* reproduites en facsimilé et traduites en français, avec une introduction et trois photographies (Paris: Gauthier-Villars, 1956), p. 114.

[34]See P. Carus, "Professor Mach and His Work," *Monist* 21 (1911), p. 33.

[35]*Where Is Science Going?,* translated by J. Murphy (New York: Norton, 1932), p. 214.

[36]Bohr's advocacy of complementarity is too well known to be documented here. What is less known is that he was, from his student days on, an admirer of H. Höffding, the Danish philosopher, who independently of W. James proposed a philosophy of pluralism. Regardless of their intrinsic merits, complementarity and pluralism are certainly efforts to overcome reductionism.

[37]As particularly evident in Bohr's life-long insistence that biology was not reducible to physics.

[38]"The Prospect for Intelligence," *Yale Review* 34 (1945), p. 450.

[39]*New Pathways in Science* (Cambridge: University Press, 1934), p. 317.

[40]*The Freedom of Man* (New Haven, Conn.: Yale University Press, 1935), p. 26.

[41]*The Function of Reason* (Princeton: Princeton University Press, 1929), p. 12.

[42]"The Departure from Classical Thought in Modern Physics," in P. A. Schilpp (ed.), *Albert Einstein: Philosopher-Scientist* (Evanston, Ill.: Library of Living Philosophers, 1949), p. 196. A very similar view was expressed by E. Schrödinger in his *My View of the World,* translated from the German by Cecily Hastings (Cambridge: University Press, 1964), pp. 7-8. Since the "previous fifteen years" mentioned by Heitler coincide with the rise of Nazism and World War II, reference should be made to the crucial role played in the formation of Nazi ideology by the physicalist interpretation of biology as championed by Ernst Haeckel and the German Monist League. This role is amply documented in D. Casman, *The Scientific Origins of National Socialism* (New York: American Elsevier Publishing Co., 1971). The popularity of the same monistic physicalism among officers of the German General Staff at German Great Headquarters in Occupied France was observed by the American biologist, V. L. Kellogg, through personal conversations. See his *Human Life as the Biologist Sees It* (New York: H. Holt & Company, 1922), p. 51. But as Bernard Shaw wrote in the Preface of his *Heartbreak House* in the wake of World War I, it was the philosophers of England who first taught that salvation was to come through Darwinist science: "We taught Prussia this religion; and Prussia bettered our instruction so effectively that we presently found ourselves confronted with the necessity of destroying Prussia to prevent Prussia

destroying us. And that has just ended in each destroying the other to an extent doubtfully reparable in our time" (London: Constable & Co., 1919, p. xiii).

[43]Reported in the *New York Times,* April 17, 1966, p. E2, col. 3.

[44]Text of address printed in *New York Herald Tribune,* April 2, 1961, Section 2, p. 3.

[45]"Science Pauses," *Fortune* 71 (May 1965), p. 116.

[46]Those papers constitute the Summer 1974 issue of *Daedalus.*

[47]I plan to show this by an analysis of three essays, whose authors, well-known astronomers, predicted around 1850, 1900, and 1950, the progress of astronomy during the next one or two generations. The discrepancy between predictions and the actual course was in all three cases astonishing.

[48]An excellent account of it is E. P. Polten, *Critique of the Psycho-physical Identity Theory* (Paris: Mouton, 1972).

[49]Any knowledge, however seminal, can be deprived of much of its meaning by such a procedure. Very likely there would be no science, as we know it today, if Newton had found it meaningless to know the reason why the elliptical orbits as established by Kepler were so peculiarly elliptical. The same remark could be made of practically any advance in all branches of science.

[50]One such humanist is Theodore Roszak. See his paper, "The Monster and the Titan: Science, Knowledge, and Gnosis," in *Daedalus* (Summer 1974), pp. 17-32.

[51]See S. Weinberg's reply to Roszak's appeal in *Daedalus,* pp. 33-46.

8

The Role of Faith in Physics

A little over seventy years ago, in 1896, the founder of psychology in America, William James, spoke before the philosophical clubs of Yale and Brown. The title of his still-famous lecture was "The Will To Believe." Its topic, as James noted with tongue in cheek, was hardly in line with what he called "Harvard freethinking and indifference."[1] In fact, a year later, when sending his lecture to print, he felt the need to explain why he had spoken of faith to an academic audience. He knew that according to most of his colleagues modern conditions required not stronger beliefs but a keener sense of doubt and criticism. Yet James did not consider it "a misuse of opportunity" on his part to emphasize the role of faith before a gathering of scholars. He admitted that credulous crowds needed to be exposed to what he called "the northwest wind of science." For intellectuals, however, he had the following diagnosis: "Academic audiences, fed already on science, have a very

This paper was originally presented as a lecture at Kansas State University, February 16, 1967. Reprinted with permission from *Zygon* 2 (1967), pp. 187-202.

different need.''[2] What they needed, according to him, was the will to believe.

It is rather a reassuring symptom that, today, academic circles suffer much less from what James called "a paralysis of their native capacity for faith."[3] The recognition is growing strong that faith, or belief, forms the ultimate foundation of the certainty of every knowledge.[4] Such is certainly the case in the field of physics. Leading physicists voice with ever greater emphasis the conviction that faith plays an indispensable role in their search for new discoveries. Their awareness is steadily growing that historic breakthroughs in physics are as much the product of a trusting faith in nature as of a critical analysis of the facts of nature. Most important, leading physicists of today know all too well that the products of science will ruin mankind unless science fosters man's faith in himself and in his goals.

In speaking about faith, one touches on a delicate subject that needs clarification, especially when related to the science of physics. No one in his right mind will have any use for a faith as defined by a schoolboy: "Faith is when you believe something that you know isn't true." Clearly, to believe in something because it is absurd would be even worse than to believe blindly, which is bad enough. One may indeed go along with the dictum of T. H. Huxley who called "blind faith the one unpardonable sin."[5] Where Huxley, however, cannot be followed, is in looking with suspicion on faith in general. Faith can, of course, be blind, but so can unbelief, and Huxley himself was blinded by a false image of science very fashionable in his day. In 1866, when Huxley made his statement, physics seemed to approach rapidly its final and perfect stage. In 1871, Lord Kelvin told the British Association that the successes of the kinetic theory of gases pointed to an early completion of an all-inclusive, definitive physical theory.[6] Two decades later, another prominent British physicist, Oliver Lodge, interpreted the success of Maxwell's electromagnetic theory in the same sanguine way. As Lodge put it: "The present is an epoch of astounding activity in physical science. Progress is a thing of months and weeks, almost of days. The long line of isolated ripples of past discovery seem blending into a mighty wave, on the crest of which one begins to discern some oncoming magnificent generalization."[7]

Neither Oliver Lodge, nor Lord Kelvin, nor Huxley guessed that, instead of a major and final triumph, agonizing discoveries were in store for physics. Discoveries were to come that played havoc with apparently absolute tenets in physics. The last decade of the nineteenth century saw the discovery of radioactivity and of X-rays. Finally, only a short three weeks before the century was out, there came Planck's announcement of the concept of the quantum of energy. The concept, as all students of physics know, stood in fundamental opposition to some basic tenets of classical physics. The concept of quantum contradicted the principle of continuity, or endless divisibility of matter, and it also seems to contradict the principle of strict, physical causality. Abandoning those principles seemed equivalent to abandoning the conviction that nature itself was orderly and intelligible. Planck himself was beset with the most serious misgivings. As a matter of fact, he explored every possible avenue to find fault with his famous derivation of the formula of energy distribution of black-body radiation.

But the concept of quantum could not be evaded. And what an ominous concept it was. It seemed to suggest that, if nature was orderly, its orderliness was beyond the reach of classical physics. But was there at that time any physics other than the classical? In the context of the times, all this seemed to mean that the orderliness of nature could not be grasped by science. As a result, the concept of quantum presented physics with a tremendous challenge. The challenge was the challenge of faith. It called for a step in the dark; it called for a step beyond the science of the day into a mysterious new land of inquiry. It was a challenge that demanded faith in the absolute orderliness of nature regardless of whether the best of science was up to it or not. Such at least was the situation as it appeared to Planck himself. To live with such a situation, to cope with it and to master it, became for him the most momentous experience of his life. It was this experience that prompted his statement of faith, which is worth being quoted in full: "Science demands also the believing spirit. Anybody who has been seriously engaged in scientific work of any kind realizes that over the entrance to the gates of the temple of science are written the words: *Ye must have faith*. It is a quality which the scientist cannot dispense with."[8]

Quantum theory is one of the two main pillars of modern physics. The other is the theory of relativity. These two theories are still unrelated. Today the so-called Unified Theory is but a dream, not a reality. There was, however, a basic common ground in the think-

ing of the authors of those two theories. Albert Einstein, the principal originator of the theory of relativity, was just as emphatic as
Planck was in stressing the importance of faith in the work of the
scientist. This is easy to understand. Relativity, no less than quantum theory, demanded an entirely new outlook on nature. The
acceptance of relativity meant the abandonment of absolute space
and time. In their place came a space and time defined in terms of
the frame of reference of the observer. No wonder that idealist
philosophers saw in relativity a vindication of their claim that the
order in nature was merely a subjective construct of the mind. Such
were not, however, Einstein's views. For him, relativity meant
rather the conviction that the laws of nature are always and everywhere the same, regardless of the frame of reference one may
choose. He viewed the constancy of the speed of light as an
absolute, primordial fact of nature that existed, with the rest of
nature, independently of the thinking mind. Furthermore, he
insisted that the scientist must have full confidence in the objective
existence of nature. "Belief," he wrote, "in an external world,
independent of the perceiving subject, is the basis of all natural
science."[9] It was the same idea that he articulated in greater detail
in his analysis of the history of physics written jointly with Leopold
Infeld. "Without the belief that it is possible to grasp the reality
with our theoretical constructions, without the belief in the inner
harmony of our world, there could be no science. This belief is and
always will remain the fundamental motive for all scientific creation."[10] To Einstein, the nature of this faith was such as to put it
into the sphere of religious beliefs. As he emphatically argued the
point, the man of science needed no less than a "profound faith" to
secure for himself the assurance that "the regulations valid for the
world of existence are rational, that is comprehensible to reason."
A scientist without that faith was simply beyond his comprehension. Clearly, such a disclosure of his thoughts had to come from
the deepest recesses of his convictions. The measure of that depth
can be best gauged in his most famous aphorism: "Science without
religion is lame, religion without science is blind."[11]

Next to quantum theory and the theory of relativity, the most
outstanding creation of twentieth-century theoretical physics is
Eddington's "Fundamental Theory." Its purpose was possibly the
most ambitious ever offered in the history of science. In substance,
Eddington tried to derive from purely epistemological considera-

tions the basic structure and fundamental laws of the universe. Thus he claimed to have established on a priori grounds that the total number of protons in the universe was of the order of 10^{79}. Eddington's ideas did not produce many disciples; yet even his most severe critics expressed their admiration for his bold efforts. At the basis of that intellectual boldness there stood an extraordinary measure of faith—faith in the orderliness of nature, and faith in the ability of the inquiring mind. Or as Eddington put it: "Reasoning leads us from premises to conclusions; it cannot start without premises; . . . we must believe that we have an inner sense of values which guides us as to what is to be heeded, otherwise we cannot start on our survey even of the physical world. . . . At the very beginning there is something which might be described as an act of faith—a belief that what our eyes have to show us is significant."[12] Long would be the list of twentieth-century physicists who spoke in the same vein. Let it suffice here to recall only a few outstanding cases. First, Heisenberg, whose indeterminacy principle showed the full depth of Planck's quantum theory. He spoke of faith as the perennial mainspring of scientific work.[13] Willem De Sitter, one of the original proponents of relativistic cosmological models, also found it important to stress that "without a solid faith in the existence of order and law no science is possible." Moreover, he was also very explicit in stating that such belief, forming the basis of science, "is not a scientific theory." It is not derived, he insisted, from science, but rather "it is prescientific, being rooted much deeper in our consciousness than science, it is what makes science possible."[14]

By referring to the concept of the possibility of science, De Sitter touched upon a point that deserves to be discussed in some detail. Most immediately, the expression "possibility of science" refers to that historic event known as the birth of science. More of that later. But the expression "possibility of science" refers also to that series of options which runs unbroken throughout the entire history of science. Of this, physicists working in the forefront of physics are fully aware. They are the ones who stand on the borderlines of the unknown. For them, the possibility of science implies a constant renewal of their faith in the orderliness of nature. The best illustration of this can be gathered from a quick glance at what goes on in high-energy physics, or the search for fundamental particles. It is a bewildering field. Hardly a month passes today without the discovery of a new particle, or resonance, or whatever name you may prefer. Theories trying to systematize those particles are succeeding one another with astonishing rapidity. The reason for this lies in the now historic pattern: each major advance in accelerator construction has brought into view new, unsuspected particles. As a physicist put it, high-energy physics seems to be caught up in an infernal race.[15]

The expression "infernal race" was well chosen from the psychological viewpoint at least. In such a race there is hardly any room for certainty or relaxation. Today, physicists think back with embarrassment to times when the last layer of matter was believed to be within reach. In our century the opening decade, the early thirties, and the fifties were such times. Thus in the early thirties the proton, neutron, and the electron were believed to have formed the fundamental system of particles. In the fifties most physicists believed that nature was built on a system of some thirty-four fundamental particles. Today, it is admitted that the best established property of fundamental particles is that none of them is fundamental. In one word, the final layer of matter appears to be farther away than ever. Recently, at the February, 1967, meeting of the American Physical Society, its president, Professor J. A. Wheeler of Princeton University, took the view that the ultimate layer of matter might be located in a practically never-never land, at the level of the so-called Planck distance, which is of the order of 10^{-33}cm.[16] How soon science will edge down to that level is anybody's guess. Perhaps in a hundred years. Even so it will be an extraordinary achievement. After all, during the last half-century, science only managed to move from the atom (10^{-8}cm.) to the neighborhood of 10^{-13}cm. This great advance covered only five orders of magnitude. Between the nucleus and the realm of Planck's distance there are, however, twenty orders of magnitude. In addition, one should not forget that the smaller a spatial magnitude is, the greater energy is required for its exploration. Whether energies necessary for the investigation of the realm of Planck's distance shall ever be available is a moot question. Furthermore, can science be assured that upon reaching that realm it would find exactly what it looked for? Very likely not. Clearly such is not a comforting outlook. It certainly gives no one the right to make easy predictions. Still the work of research must go on. And it is well to remember that its ultimate sustaining force is faith. Or to hear a prominent physicist, the late director of the Institute for Advanced Study, Robert Oppenheimer, state it: "We cannot make much progress without a faith that in this bewildering field of human experience [particle research], which is so new and so much more complicated than we thought even five years ago, there is a unique and necessary order; not an order that we can see without experience, not an order that we can tell a priori, but an order which means that the parts fit into a whole and that the whole requires the parts."[17]

Ten years have passed since Oppenheimer made this statement of faith. Those ten years were an era of feverish research, yet none of the results diminished either the beauty or the truth of his words. No physicist can tell us today what are the true parts of the ultimate system of particles; yet, all believe firmly in the existence of such a system. This faith of theirs is not an easy one. After all, they are everyday witnesses to the fact that assuredly stable particles turn out to be subject to decay. Thus the concept of finality or definiteness has taken on for the modern physicist a meaning wholly different from its obvious meaning. Finality is to be taken today in physics as largely provisional. It ought to be most puzzling for the modern physicist to find that it is his own tools that time and again deprive him of apparently firmly established grounds. These tools are the tools of precision. They both confirm and undo theories, and keep physics in a dynamic flux never experienced before. These tools create as many problems as they provide solutions. For all that, the physicist must retain his confidence in the double-edged sword of precision, which keeps opening up before him strange, perplexing worlds. In using the tools of precision, all physicists are sustained by faith. It holds of all of them what was true of Albert A. Michelson, a wizard of precision in measurements and the first American to receive the Nobel Prize in physics. As Millikan, another Nobel laureate, said of Michelson: "He merely felt in his bones or knew in his soul, or had faith to believe that accurate knowledge was important."[18]

It was more than forty years ago that Millikan uttered these words. In American science and scientific philosophy, the thirties were still an era dominated by cliché accounts of the history of science. It was an era that accepted without further ado the slogan that physics consisted solely in correlating data of observations and experiments. The word "faith" was an ugly word for most of those who in those years and until very recently posed as the supreme interpreters of science and were accepted as such. I have in mind the neopositivists and the operationalists. There is, of course, much that can be said in favor of operationalism and of logical positivism. When, however, taken as the fundamental and exclusive theories of science, they display a serious shortcoming. Operationalism and logical positivism do not square with the facts of scientific creativity. In our times this was emphasized by such creative personalities of physical science as Einstein, Born, Schrödinger, and many others.[19] It was in fact in the wake of his discovery of wave-mechanics that Schrödinger decried "that cold clutch of dreary emptiness" which exudes from the definition of scientific work as given by positivism: a description of the facts, with the maximum of completeness and the maximum economy of thought. Scientists sufficiently honest with themselves, Schrödinger added, would admit that "to have *only* this goal before one's eyes would not suffice to keep the work of research going forward in any field whatsoever."[20]

Much less could the positivistic concept of science give start to the scientific endeavor itself. No wonder that the very start, the birth of science, has not become a favorite topic with positivist historians of science. Indeed, there can be no satisfactory explanation for it within a framework that frowns on the mental attitude called faith. Within the positivist framework it must remain an insoluble puzzle why science was born in the Western world and not in China or India or among the Maya and the Aztecs. The birth of science was, of course, a rather long process. Its beginnings credit the marvelous insights of the Greek mind. As Einstein once noted: "In my opinion one has not to be astonished that the Chinese sages have not made these steps [the major discoveries of Greek science]. The astonishing thing is that these discoveries were made at all."[21] Still for all its achievements, ancient Greek science is not without a grave puzzle. That puzzle derives from the fact that Greek science remained a half-way house. It failed to recognize the crucial role of systematic experiments. It proved itself wholly powerless to come to grips with the quantitative analysis of motion.

It is a fact of scientific history that man needed faith to overcome these hurdles and to bring science to a full birth. It is a fact of scientific history that the birth of modern science took place in a cultural ambiance wholly permeated by belief in dogmas. Foremost of these was the Christian tenet about a personal, rational Creator of the universe. Our century was reminded of this by Whitehead in his Lowell Lectures of 1925, published under the title, *Science and the Modern World.* To millions of readers of that book it came as a revelation that, contrary to the claims of positivism, science does not owe its origin to the rejection of religious beliefs. Instead, as Whitehead told his readers, they had to look for the birth of science in the staunch belief of the Middle Ages. Foremost in this respect was, according to him, the medieval insistence on the rationality of the Creator. Whitehead also emphasized that belief in the dogma of creation had to be shared by a whole culture throughout several generations. Only such communal experience and conviction could produce what Whitehead called a tone of thought, a climate of intellectual confidence and courage.[22] This in turn gave rise to the scientific enterprise and determination to look for rationality in every process of nature.

In Whitehead's classic discourse, only one point was missing. He should have called his listeners' and his readers' attention to the fact that what he said was not a more or less subjective version of history. He should have told them that his ideas were but the echo of those men of science who witnessed the birth of science three to six centuries ago. Thus references to the Creator are explicit in the great medieval forerunners of modern science, such as Oresme and Buridan. Their statements were further elaborated by such theoreticians of sixteenth-century science as Descartes, Bacon, and Galileo. Bacon's writings in this respect are especially instructive. Not a particularly original thinker, Bacon had an uncanny sense of gathering the best that was available in his time. He also had the skill to elaborate on it with great persuasiveness. Most of all, he said what his contemporaries wanted to hear. They wanted to hear, for instance, why Greek science came to a standstill. For the failures of the Greeks, Bacon laid the blame on the pantheistic features of their religious views.[23] It was pantheism that put the theological seal on the Greeks' preference for viewing the world as an organism, or a huge animal. For them, each portion of the world was full of volitions closely paralleling human strivings and aspirations. They discussed the fall of stones, the rise of fire, the motion of the stars in the same breath with the motion of animals. For them, man was but a tiny organism wholly subject to the countless volitions animating the whole cosmos. Obviously, such an outlook could not generate a sustained confidence in ever deciphering, let alone mastering, the whims and movements of that great animal, the entire universe.[24]

On the sad failure of Greek science, an unexpected light is thrown by recent investigations of Chinese scientific history. What I have in mind is the conclusion of J. Needham, the distinguished author of the most monumental study of the history of Chinese science ever published in the West. A Marxist, Needham looked in various socioeconomical factors for the likely cause of the failure of the Chinese to invent science, so to speak. As is well known, ancient and medieval Chinese, though very proficient in practical inventions, such as rockets and compasses, failed to formulate one single law of physics. As might be expected, Needham laid part of the blame on medieval Chinese feudalism and other so-called reactionary factors. Yet, according to Needham, the fundamental reason for the scientific failure of the Chinese lay somewhere else. He had to admit that the basic cause of that failure pointed to theology. More specifically, he called attention to the early loss in Chinese religious thought of the belief in a personal rational Creator. With the loss of that belief was also lost the faith, the confidence of the Chinese in the ultimate rationality of the universe. To quote Needham, "Among the Chinese there was no conviction that rational personal beings would be able to spell out in their lesser earthly language the divine code of laws which the Creator had decreed aforetime."[25]

It was not, therefore, a freak happening of history that science was born in a Europe that was living through its centuries of faith. It was a Europe where those lived and worked who looked upon the world as the product of a most rational Creator and looked upon themselves as the stewards of their Father's handiwork. Theirs was not a blind faith, and happily for them. For the twist of history thrust upon them the whole Greek scientific corpus within the short span of two generations. What hit them was nothing short of an intellectual deluge. All of a sudden they were challenged by the dazzling scientific works of a Euclid, of a Ptolemy, and of an Aristotle. Some of the passages they could not translate, let alone understand. But they did not panic. Instead, they read those books with eager enthusiasm, notwithstanding the fact that Rome at one time put a ban even on the works of Aristotle. The enthusiasm of the medievals is easy to understand. They believed themselves to be

children of an all-powerful, all-reasonable, all-good Creator. Consequently, they had to be enthusiastically confident in the final outcome of their newborn quest for scientific knowledge.

The quest of science has seen many triumphs and many agonies.
They usually went hand in hand and evidenced equally well the role
of faith for science. The first major triumph was Copernicus' outline of the planetary order. He was far from proving definitely the
heliocentric proposition. But what he lacked in physical proofs, he
amply supplemented with his faith in nature. From his belief that
nature was the handiwork of the Creator, he readily concluded that
nature was simple. His system of the planets, it is well to recall,
gave no better prediction of the motion of planets than did
Ptolemy's; the most attractive proof of Copernicus lay in the geometrical simplicity of the new ordering of the planets. It was a bold
view, and he clung to it though its consequences flew in the face of
everybody's daily experience. Positivists of all times may shake
their heads in disbelief, but Galileo, whom they consider the father
of experimental method, praised Copernicus precisely for what he
did: for staying with his belief at the price of committing rape of his
senses.[26]

These words of Galileo are not without some irony. Though he
praised the faith of Copernicus, he did his best to conceal the fact
that much of what he claimed in the *Dialogues* was still largely a
matter of faith. He passed over in silence the fact that his
unbounded admiration for geometry was in effect a loud profession
of his faith in the geometrical ordering of nature. Mystic as he was,
Galileo frowned on anything savoring of mysticism, and soon developed a dislike for Kepler, an unabashed mystic. The loser was
Galileo. Had he referred in his *Dialogues* to Kepler's Laws, he
might have considerably strengthened his cause. Also, his conflict
with some churchmen might have taken a different course if it had
been recognized that there is a role for faith in science and that
theology does not operate by faith alone.

When the clash came to a head, Kepler was already dead. Perhaps he could have testified that his three laws were the outcome of
tedious computations as well as of his firm faith in the mathematical orderliness of the universe. For this, no one gave him
greater credit than Max Planck. In fact, Planck found a startling
analogy between his case and Kepler's struggles. In Planck's case,
the data of blackbody radiation were available to a great number of

his colleagues. Yet, only one, Planck himself, perceived the true pattern underlying those data. And Planck was not ashamed to ascribe that success to his faith. Now, as Planck analyzed Kepler's case, both Tycho Brahe and Kepler were in possession of the same data of planetary motions. Yet, only Kepler found their true correlation. The answer to this could not be clearer to Planck. As he put it, Tycho did not have what Kepler did possess: scientific faith.[27]

That scientific faith is in evidence in all major breakthroughs and principal tenets of science. Men of science had believed in the inverse square law of gravitation long before its truth was demonstrated. Maupertuis had believed in his law of least action years before he formulated it with enough clarity. The law earned him the ridicule of the rationalist Voltaire, who decried it as credulous metaphysics. Yet, ultimately, it was Maupertuis' faith that proved victorious. It received its due praise when Helmholtz discussed the law of least action in 1884 before the Berlin Academy of Sciences. There Helmholtz traced the origin of the law to Maupertuis' belief in the uniformity of nature and in the human mind's ability to find the true form of that uniformity.

That Helmholtz saw Maupertuis' efforts in this light is understandable. Faith was the mainspring of his efforts to have the law of the conservation of energy recognized. His was not an easy struggle. His now classic paper, "On the Conservation of Force," was rejected by the leading German physical review. In the long run, however, the faith of Helmholtz prevailed. And so did the faith of Faraday and of other great physicists who worked on proving that all forces of nature are interconnected. In the case of electricity and magnetism, Faraday's was a complete success. On the other hand, only failures accompanied his lifelong efforts to find a correlation between electromagnetism and gravitation. For all that, the entries in his notebooks on the subject never showed the slightest trace of wavering. All the failures, he remarked, "do not shake my strong feeling of the existence of a relation between gravity and electricity." One of his papers on the subject refers to "the full conviction" and, again, "to the same deep conviction" that animated his search for a connection between gravity and electricity. To follow the promptings of that "strong feeling" was in his view a most sacred scientific duty. The contrary course, that is, to leave the problem untouched seemed to him equivalent to abandoning faith in nature or, to quote his words: "to rest content with darkness and to worship an idol."[28]

Fortunately for science, Faraday's faith, or "strong feeling," or "full conviction" in the interconnectedness of the forces of nature is as alive as ever. Witness Einstein's thirty years of search for a Unified Theory; witness the efforts to find a connection between the nuclear force and the force of the so-called weak interactions. Or witness the rather recent competition for the best essay on the possibility of gravitational shielding.[29] The idea underlying the competition was that, if there is a shielding against electrical forces, the same should also be true of gravitation. Faraday, I am sure, would have found to his liking a contest of this type and most likely would have participated in it with a lengthy paper. He would have also found that no less than in his time, physics in the 1960s is still supported both by evidence and by faith—by faith in the interconnectedness of the parts of nature; by faith in the intelligibility of nature; by faith in its simplicity, in its uniformity, and in its symmetry.

Intelligibility, simplicity, and uniformity of nature are concepts that are rarely reflected upon. They are like the air we breathe, they are taken for granted. All too often they are treated as self-evident notions that need no further scrutiny. Yet, when scrutinized with no reference to the scientist's faith in them, what remains of them? In a positivist framework of explanation they are reduced to formulas of convenience devoid of that absolute certainty with which the scientist espouses them. For once the principles of positivism are consistently applied, one cannot even have absolute certainty about one's own existence. Or as H. Reichenbach, a leading positivist philosopher of science, claimed: "We have no absolutely conclusive evidence that there is a physical world, and we have no absolutely conclusive evidence either that we exist."[30]

A long comment could be made about such a position, but let me confine myself to the most obvious. Whatever the validity of Reichenbach's claim, the scientist needs in his work an unconditional and complete trust or conviction in his own existence, in the existence of nature, and in its simplicity, orderliness, and intelligibility. On such points, the scientist can entertain no misgivings, no futile sophistry, no wholesale doubts, no endless questioning. The scientist must go beyond the set of evidences available to him and must assert that nature in its ultimate foundations is absolutely simple and perfectly ordered.

Of course the scientist's evidence of the simplicity and orderliness of nature is much more extensive than that available to the ordinary layman. Yet, even the scientist's glimpse of that orderliness is far from being exhaustive. The condition of the scientist is therefore much the same as that of the man of religion. Religious faith, like the faith of the scientist, has its set of evidences. Religious faith is not a blind faith.[31] Yet, numerous as its evidences might be, they do not form a complete, exhaustive set. Those evidences, like the evidences of science, are rather a prompting toward espousing propositions that imply unconditional affirmation and absolute commitment. It is through such commitment that the man of science grasps the simplicity and order of nature, and it is through a similar commitment that the man of religion grasps the spiritual and moral dimensions.[32]

This short outline of the analogy between scientific and religious faith was not prompted by some hidden aim of proselytizing. The meaning and purpose of the analogy is far deeper. It is my conviction that the recognition of that analogy is of paramount importance if a major tragedy of our culture is to be overcome. That tragedy is the split of our culture in two sections. Today, intellectuals are clustered in two camps; they are either humanists or scientists. They speak different languages, they hardly communicate with each other, and consider each other's problems as largely irrelevant.

Much has been said about that cultural split, and well before C. P. Snow came up with the now famous phrase, "two cultures." The tremendous response given to his work, *The Two Cultures,* is in itself evidence that the cultural split is a reality and a dangerous one. For that split, Snow laid much of the blame at the door of the humanists. It was in line with this that Snow prescribed his medicine for the restoration of the cultural unity. The medicine consisted in compulsory science courses, and a fair number of them to be imposed on students of humanities. I would not dispute that today students of humanities should do their best to become very familiar with science. Yet, just as important as the science one knows is one's familiarity with the foundations of the scientific quest. A careful study of those foundations will show that the sciences and the humanities have at their bases some remarkably common mental attitudes. One of them, and possibly the foremost, is the attitude of faith.

I know that the word "faith" is loaded with too many connotations to be readily acceptable to many. If so, I am not reluctant to look for a substitute expression. To me, a most appealing one was coined by none other than David Hume, hardly a friend of intellectual faith. He preferred to speak of faith as a "kind of firm and solid feeling." Regardless of Hume's philosophical outlook, I find the expression to be one that perfectly suits our purpose here. A full recovery of that "firm and solid feeling" by today's intellectuals would greatly help to forestall the threat posed to human values by an unbridled technologization of life. Today the evaluation of man is shifting more and more toward the quantitative aspects. Calipers, slide rules, statistics, and computers are being used in areas where they can never come even remotely close to the heart of the matter. For numbers, equations, and tools, however precise, can never touch on the very core of man and on his faith or, if you wish, on his strong and firm feelings. Computers may be said to do thinking, but only man feels in the sense of having faith. Therein lies man's basic dignity and also his most perennial need. The scientist is no exception to that rule. The man of science, like all his fellowmen, lives by faith and ultimately makes his progress in virtue of his faith.

[1]William James, *The Will To Believe and Other Essays in Popular Philosophy* (New York: Longmans, Green & Co., 1897), p. 1.
[2]Ibid., p. x.
[3]Ibid.
[4]The role of faith in scientific inquiry is rich in aspects, some of which have been given illuminating treatment in recent literature. Foremost to mention is the work by M. Polanyi, *Science, Faith and Society* (London: Oxford University Press, 1946; reprinted with a new Introduction by the author: University of Chicago Press, 1964). Some valuable contributions to the subject were made by noted physicists, such as H. Margenau, *Open Vistas: Philosophical Perspectives of Modern Science* (New Haven, Con.: Yale University Press, 1961), pp. 73-76; K. Lonsdale, *I Believe: The Eighteenth Arthur Stanley Eddington Memorial Lecture, 6 November 1964* (Cambridge: University Press, 1964); H. K. Schilling, *Science and Religion: An Interpretation of Two Communities* (New York: Charles Scribner's Sons, 1962).
[5]T. H. Huxley, "On the Advisableness of Improving Natural Knowledge,"

in *Method and Results: Essays* (New York: D. Appleton & Co., 1894), p. 40.

[6]See Lord Kelvin's presidential address in *Report of the Forty-first Meeting of the British Association for the Advancement of Science* (held at Edinburgh in August, 1871) (London: John Murray, 1872), p. xciii.

[7]Oliver Lodge, *Modern Views of Electricity* (London, 1889), pp. 382-83.

[8]Max Planck, *Where Is Science Going?* translated by J. Murphy (New York: W. W. Norton & Co., 1932), p. 214.

[9]Albert Einstein, "Clerk Maxwell's Influence on the Development of the Conception of Physical Reality" (1931), in *The World as I See It* (New York: Covici, 1934), p. 60.

[10]Albert Einstein and Leopold Infeld, *The Evolution of Physics* (New York: Simon and Schuster, 1938), pp. 312-13.

[11]Albert Einstein, "Address to the Conference on Science, Philosophy, and Religion" (1940), in *Out of My Later Years* (New York: Philosophical Library, 1950), p. 26.

[12]A. S. Eddington, *Science and the Unseen World: Swarthmore Lecture, 1929* (New York: Macmillan Co., 1930), pp. 73-74.

[13]W. Heisenberg, "A Scientist's Case for the Classics," *Harper's Magazine,* CCXVI (May, 1958), p. 29.

[14]Willem De Sitter, *Kosmos* (Cambridge, Mass.: Harvard University Press, 1932), p. 10.

[15]L. Brillouin, *Scientific Uncertainty, and Information* (New York: Academic Press, 1964), p. 41.

[16]On Wheeler's lecture, see W. Sullivan's report in the *New York Times,* February 5, 1967, sec. E, p. 5, cols. 3-5.

[17]Robert Oppenheimer, *The Constitution of Matter* (Eugene: Oregon State System of Higher Education, 1956), p. 37.

[18]R. A. Millikan, *Science and the New Civilization* (New York: Charles Scribner's Sons, 1930), p. 164.

[19]On this point, see my work, *The Relevance of Physics* (Chicago: University of Chicago Press, 1966), pp. 479-80.

[20]Erwin Schrödinger, *My View of the World,* translated by C. Hastings (Cambridge: University Press, 1964), pp. 3-4.

[21]Einstein, in a letter of April 23, 1953, to Mr. J. E. Switzer; see D. J. de Solla Price, *Science since Babylon* (New Haven, Conn.: Yale University Press, 1961), p. 15.

[22]Alfred N. Whitehead, *Science and the Modern World* (New York: Macmillan Co., 1926), pp. 18-19. For a very valuable discussion of the import of the Christian doctrine of creation, see L. Gilkey, *Maker of Heaven and Earth: The Christian Doctrine of Creation in the Light of Modern Knowledge* (1959) (Doubleday Anchor Book reprint; Garden City, N.Y.: Doubleday & Co., 1965). Concerning the Christian origins of modern science, Gilkey's discussion needs updating. Modern historical research has clearly

shown those origins to be medieval, a point that was ignored by Gilkey's principal source on this point, several articles by M. Foster, published in *Mind* between 1934 and 1936.

[23]Bacon, *Of the Dignity and Advancement of Learning,* Book 3, chap. iv, in *The Works of Francis Bacon,* edited by J. Spedding, R. L. Ellis, and D. D. Heath (new ed.; London, 1870), vol. IV, p. 365.

[24]On the impotence of the organismic concept of the physical world, see my *Relevance of Physics* (n. 19 above), chap. i.

[25]J. Needham, *Science and Civilization in China, II: History of Scientific Thought* (Cambridge: University Press, 1956), p. 581.

[26]Galileo, *Dialogue concerning the Two Chief World Systems,* translated by Stillman Drake (Berkeley: University of California Press, 1953), p. 328.

[27]Max Planck, *Where Is Science Going?* (n. 8 above), p. 214; see also his *The Philosophy of Physics,* translated by W. H. Johnston (New York: W. W. Norton & Co., 1936), pp. 122-23.

[28]For a convenient source on these statements of Faraday, see H. Bence-Jones, *The Life and Letters of Faraday* (Philadelphia: J. B. Lippincott Co., 1870), vol. II, pp. 253, 417, 388.

[29]It formed part of a program sponsored by the Gravity Research Foundation that for a number of years has awarded prizes to outstanding essays on various aspects of the problem of gravity.

[30]H. Reichenbach, *The Rise of Scientific Philosophy* (Berkeley: University of California Press, 1951), p. 268.

[31]Thus N. Wiener took pains to emphasize that the faith needed in scientific work has nothing in common with religious faith which he described as a set of dogmas imposed from outside (*The Human Use of Human Beings* [reprinted by Doubleday & Co., n.d.], p. 193). Religious faith was therefore rejected by Wiener as "no faith." Such high-handed, if not superficial, handling of the concept of religious faith proves only one thing. A scientist, however eminent, may easily dispense, when discussing topics outside his field, with the elementary scientific duty of securing for himself a fair measure of proper information in the matter.

[32]There are, of course, differences between the attitudes of faith as acted out within the religious and the scientific framework, respectively. Those differences mainly derive from the role played by revelation and authority as normative factors within the community of the faithful. The rise and growing influence of science was most beneficial in reminding theologians and churchmen that those normative factors are restricted to moral and supernatural considerations and can never play a heuristic role in man's search for the regularities of the processes of nature.

9

Theological Aspects of
Creative Science

When Whitehead put on the seventeenth century that happy label, the century of genius, no explanation was needed about the kind of genius he had in mind. Kepler, Galileo, Descartes, Harvey, Boyle, Huygens, Leibniz, and Newton showed their genius as men of science. They created science, or rather they raised science to a stage where it seemed to be possessed of an undying vigor. This truly creative achievement had several characteristics of which the most conspicuous was the close alliance of natural science with natural theology. Kepler spoke of Copernican scientists as priests officiating around the altar of the Creator.[1] Newton in turn made no secret of his pleasure that Bentley had found the *Principia* to be a storehouse of pointers toward the Maker and Creator of all.

This paper is based on a lecture given at Princeton University on Feb. 20, 1975, in commemoration of the centenary of Albert Schweitzer's birth. Reprinted with permission from *Creation, Christ and Culture: Studies in Honor of T. F. Torrance,* ed. R. W. A. McKinney (Edinburgh: T. & T. Clark, 1976), pp. 149-66.

Between Kepler and Newton the virtuosi made it a tone of thought that scientific work was, in Boyle's words, a vehicle to the "seraphick love" of God.

A hundred years later the atmosphere between science and theology was noticeably different. Far from echoing sounds of jubilation, the air was quiet, though not entirely so. Small sparks of electricity broke the silence with their crackling sound. One such spark was Voltaire's complaint about the piety of Euler, the greatest scientist of the day, a piety which Voltaire ascribed to Euler's senility. About the same time, in the early 1760s, Lambert, the self-made genius, became the butt of snide remarks as he regularly took part in communion services in the Reformed Church of Berlin. It was a city where the academicians were overawed by Voltaire, to say nothing of those many who stood in uninformed awe of the academicians. The same age saw d'Alembert become the victim of unsavory libel by Diderot, who found it intolerable that the greatest French "geometer" did not follow him on the primrose path to rank atheism. Diderot did not live to see the day when Laplace, the foremost student of d'Alembert, stood up to the First Consul, who deplored the absence of any reference to God in Laplace's account of the solar system. The witness of the dispute was none other than Herschel, who quietly commented in his diary of his visit of 1802 in Paris: "Much may be said on the subject; by joining the arguments of both we shall be led to 'nature and nature's God'."[2] But Herschel kept his belief in God largely to himself. Half a century later his son, himself an illustrious astronomer, had to defend his father's name against innuendoes of atheism.

Those innuendoes were part of an open warfare which was declared, shortly after the middle of the nineteenth century, on religion in the name of science. The names of great scientists were conspicuously absent among those in the forefront of that virulent attack. Moleschott, Vogt, Büchner, Engels, Huxley, Spencer, Littré, White, Draper are not names of scientific discoverers, but of propagandists of an interpretation of science which rested on the conviction that science and religion were in irreconcilable conflict. The means of spreading this conviction consisted in setting off the "latest" in science against antiquated accounts of theology. The victims of this strategy were many and some very illustrious. After reading some "popular" accounts of science, Einstein reached the firm conclusion at the tender age of twelve that biblical revelation had no rational foundation.[3] Whatever he wrote and said in later life about religion remained within the confines of that youthful self-instruction.

On a cursory look nothing much positive emerges when a survey is made of the relatively little written about religion by the most prominent figures of twentieth-century science.[4] Planck, with his groping for a personal God, still belonged to an older school, which, however, shied away from historical revelation. Bohr's views on religion were those of Harald Höffding, the Danish forerunner of William James. They amounted to the recognition of some purely natural aspirations in man complementing sheer rationality. In Schrödinger's Buddhism there was no room for a transcendental, personal God, let alone for His stepping into history through a specific revelation. The *Physics and Beyond* of Heisenberg contains no metaphysics worthy of that name. Its concluding note, the enthralment of a Beethoven trio, is certainly beyond physics, but not at all beyond *physis* or nature. Pascal's fervent commitment to the God of Abraham, Isaac, and Jacob, revealing himself in Jesus, had no appeal for Heisenberg.[5]

Others, like De Broglie and Dirac, kept a studied silence about religion, in accordance with the widely shared view that science alone is public knowledge, or a knowledge with objective validity, whereas religion is merely a private knowledge, that is, a respectable personal opinion at best.[6] The twentieth century is not, of course, lacking in prominent scientists with faith in a personal God and even with genuine commitment to historical, revealed Christianity. Today, as at any other time, the statistical distribution of scientists along the gamut ranging from rank disbelief to strong belief matches the distribution of other educated men along the same scale. It was even claimed by C. P. Snow that among the younger ones there were more with a penchant for religion than without.[7]

Still, the largely prevailing attitude among scientists of our times is that religion is not to be mentioned publicly and certainly not in connection with science. Open attacks on religion and God are not applauded as loudly as they were a century ago. It is largely recognized in at least the civilized parts of the globe that rude attacks are counter-productive. At any rate, religion and God seem to have been successfully eliminated from the public arena of intellectual discourse. Such is a striking contrast not only with the witness on behalf of God and Christian religion by Oersted, Ampère, Faraday, Fraunhofer, Helmholtz, Joule, Maxwell, Fizeau, Clausius, and Kelvin, a blue-ribbon list of nineteenth-century physicists, but also with the list stretching from Kepler to Newton. The rise of science to a unique level of creativity during the century of genius is an indisputable fact and so is the sound of jubilation which accompanied the advance of those geniuses from nature to nature's God. The twentieth century is certainly a match to the seventeenth as far as scientific genius is concerned. But the erstwhile jubilation is now largely echoed by a lame silence or by inept words about God and religion. Is it then still legitimate to speak about the theological aspects of creative science?

Unfortunately, the word creative has lost much of its meaning in its present-day overuse. There was a time when only God was the proper subject of the verb create. In the Old Testament the word *bara'* was reserved to an action which only God could perform and wherever Christianity made its imprint on a culture, it became part of the cultural consciousness that God alone could create. Those aware of the workings of inner logic will not be surprised by the very different use of the verb create in non-Christian cultures. Our post-Christian culture is no exception. Today it is a mark of scholarship to claim that it is not God who created man or anything, but it is man who creates his gods. Whatever the merit of such scholarship, it is certainly a sign of our cultural poverty that, in many of our schools, courses in creative writing are offered on levels where spelling and grammar still could be taught with great profit.

This abuse of the word creative, and many other examples could be mentioned, is not merely the doing of those who are usually called the humanists. The abuse is just as much present in scientific literature. A good example is *The Creation of the Universe,* one of the "musts" in the 1950s. From its second printing on it carried the warning of its author, George Gamow, that he meant creation only in the sense "of the latest creation of Parisian fashion."[8] Gamow was right. It was in that sense that he used the word. It was, of course, his privilege to use "creation" in the sense in which the world of fashion uses it. It is another matter whether truly creative scientists have ever claimed the privilege of looking at the world as if it were a mere fashion.

About the time when Gamow's book had its heyday, the rage in cosmology was the steady-state theory. It is based on the postulate that hydrogen atoms are constantly created everywhere in cosmic spaces to maintain the density of matter the same while the galaxies are receding from one another. Like Gamow, the proponents of the steady-state theory, Bondi, Gold, and Hoyle, had to explain themselves before long on the meaning of creation. As Bondi made it clear in his *Cosmology,* the creation of hydrogen atoms was a formation of matter out of nothing.[9] Such is truly a creation which, when severed from any reference to the Creator, merely begs the question. Compared with it the teaching of creative writing to practical illiterates may seem a minor self-deception.

Gamow, Bondi, Gold, and Hoyle certainly deserve to be called good scientists. Hoyle, in particular, will certainly be remembered by historians of stellar physics. Future cosmologists and especially historians of cosmology will remember Hoyle, as well as Bondi and Gold, as curiosum, in much the same way in which historians of gravitation today recall George Louis Le Sage. He was the author of a very attractive but wholly wrong explanation of gravitation by impact. With his mechanical theory of gravitation Le Sage merely gave around 1780 another application of the mechanistic creed according to which all physical influences derive from physical contact. By claiming that the universe has since eternity been the same and will always remain the same, on a large scale at least, the steady-state theorists only gave a new twist to an age-old dogma, the eternity of the world, which in the hands of Aristotle put physics and cosmology into a straitjacket for two thousand years.

As to Gamow, he certainly made some memorable contributions. Among them are his explanation of the emission of alpha particles from radioactive elements and his prediction that the start of the expansion of galaxies should produce a background radiation. With his theory of alpha tunnelling Gamow did not create quantum theory; he merely worked within his context. He did not discover a large new continent, he was merely the first to survey one of its hidden valleys. Exactly the same holds true of his prediction of the cosmic background radiation. By predicting it Gamow did not discover general relativity and its cosmology. He merely unfolded one of its many implications.

The creators of quantum theory and of general relativity were Planck and Einstein. They were the great discoverers; the many other great names following in their footsteps were the pioneering surveyors. Of such great discoverers, there were very few in science. They seem to have come in groups as science progressed. Planck and Einstein form one such group. Leucippus and Socrates would form the earliest of such groups. The next group is that of Copernicus, Kepler, Galileo and Newton. With some reservations one can add another group, Faraday, Helmholtz and Maxwell. A total of less than a dozen names. If any reduction is to be made in that list, neither Planck nor Einstein would be among those to be omitted. While the unity of forces was a much-talked-about topic some time before Faraday, to say nothing of Helmholtz and Maxwell, nobody before Planck spoke of quanta of radiation and nobody dreamed of the equivalence of accelerated inertial systems before Einstein. Again, Copernicus was not supported by any previous trend toward heliocentrism and nobody before Newton spoke of gravitation in the sense he did.

So much in defence of limiting the number of eminently creative scientists to less than a dozen, and to a mere two in this century of science explosion. Such a stringent definition of scientific creativity is certainly not in opposition to the fact that all recent efforts to probe into the psychology of creative thought[10] only strengthened its mysterious character. At any rate, such a stringent definition of "creative" and "creative science" will not lend itself to marketing children's toys under the same label. The rarity of great creative thought is not its only aspect that can be handled with some ease. Another can be studied with profit in the literature on the problem of scientific discovery. This ample literature[11] makes clear at least two features of scientific creativity. One is its considerable independence of social parameters. Sociology may explain many things; unfortunately in the hands of some of its cultivators it explains everything. But the sociology of science does not explain even such minor details as, say, the conflict of Gassendi and Descartes. Both came from the same seventeenth-century French bourgeoisie, both were Roman Catholics, and yet they became the spokesmen of radically different philosophical and physical theories. Sociology does not explain why the stiff Prussian, vaguely Christian, and socially conservative Planck, became the most perceptive supporter of Einstein, an agnostic Jew, a political radical, and very informal in his behavior. Of course, those who derive both quantum theory and general relativity from the gentle decadence of the turn of the century[12] will find it difficult to explain why the lives and personalities of both Planck and Einstein were free of symptoms of what is commonly called decadence.

In addition to being remarkably free of social parameters, scientific creativity is also elusive to psychological probings. The best that had been found about scientific creativity through psychological investigations is to be credited to Gestalt psychology. But when Gestalt psychologists, or philosophers and historians of science working with its tools,[13] state that the sudden insight marking the moment of scientific discovery is an indivisible whole, they merely state the fact. Gestalt psychology is a statement, a recognition but not an explanation of the process of intellectual perception in general and of creative perception in particular. Gestalt psychology is not so much an explanation as a reaction, a most valuable one against psychologies based on Hume's empiricism and Mach's sensationism. With Hume and Mach, the bits of sensory impressions could never really come together into that wholeness which Gestalt psychology rightly takes as the hallmark of cognition. While not an explanation, Gestalt psychology is a wholesome antidote to empiricism and sensationism in which science and scientists are created by sensory data, instead of scientists creating science with the eyes of their minds fixed on the data.

There is a third branch in present-day literature on the history and philosophy of science which sheds a useful light on the question of creative science. This literature is concerned with the scientific revolutions.[14] As in the case of the word creation, the word revolution, too, can be used in a sense which is almost a parody of its original use in the scientific context. In our times the word revolution is inseparable from the impression which compares well with the one given by fermenting grapejuice. In the process the whole surface becomes covered with unseemly foam and the air above it is filled with a heavy odor. Yet such a description of revolutions is a bit Manichean. Human nature, even those of revolutionaries, is never wholly evil. It can be terribly misguided at times, but even then it is driven by a vision about a perfect state of things. The more ferocious a revolution is, the more forceful is that vision of the ideal. All revolutions, however bloody, were preceded by the Utopian vision of social mystics. Unfortunately, before a place becomes actually called "Place de la Concorde," the guillotine has to flood it with innocent blood.

Scientific revolutions also involve, if not head chopping, at least some head hunting. Luther called Copernicus a fool, Galileo was muzzled by two Popes. Around the turn of this century Boltzmann was driven to suicide in part at least because of Mach's bitter campaign against atomism. Mach's efforts to discredit Einstein were more veiled but no less resolute. Planck, another target of Mach, drily noted that before a new theory is accepted its opponents must die out.[15] By a new theory Planck meant not some novelty but a novel vision of a perfect order of things that lies behind scientific revolutions. Copernicus called his immortal book *On the Revolutions of Celestial Orbs,* precisely because those revolutions mirrored a perfect order, the order of planets revolving in concentric orbits around the sun. The vision was partly philosophical, partly theological. It was steeped in a faith in the Creator's eternal wisdom and decree. The same is true of Kepler, Galileo and Newton. This is a fact of the historical record, and it was a very important fact for those who like Copernicus and the others created at that time the scientific revolution. Unfortunately, no justice is done to this in much of the current literature dealing with scientific revolutions.

A happier fact about that literature is the general agreement that scientific revolutions are few and far between. The very small number of scientific revolutions may sound a truism hardly worth any excitement. But in this age when consensus among scholars is so hard to find, this particular agreement should be the cause of some rejoicing. It is agreed that there have been two scientific revolutions so far. One was initiated by Copernicus and completed by Newton, the other is ascribed to Planck and Einstein. If the Greek background of science is considered, then a third may be added, the Socratic revolution. It was first outlined in the *Phaedo* whose theme Plato and Aristotle were to elaborate. Socrates advocated a world view in terms of the biological organism. The Copernican revolution changed the organism to mechanism. In the twentieth century science looks at the world as a mathematical pattern.[16]

Equally important is the fact that the shift from one world view to the other was a very conscious process. Moreover, eighteenth- and nineteenth-century physicists were fully aware of the fact that they were not creators or discoverers on a par with Newton, but merely surveyors of a new continent discovered by him. No less an authority than Lagrange said precisely this around 1800 as he remarked on the good fortune of Newton to have made the discovery which could be made only once.[17] Lagrange also showed in detail that all subsequent advances in theoretical mechanics were only the unfolding of principles laid down by Newton. At the end of the century, Mach emphasized the same in the successive editions of his *Science of Mechanics*.[18] The awareness of twentieth-century scientists of their debt to Planck and Einstein is too clear to need any illustration.

The very small number of scientific revolutions and the very small number of scientists to be credited with initiating those revolutions is worth stating partly because it constitutes a point of agreement among quarrelling scholars. Compared with those few giants in physical science there are an immense number of scientists of different stature, some very great, though not giants, many more of common stature, and an enormous number of plain dwarfs, when compared with giants. The seven dwarfs, it should be recalled, were crucial for the success of the only Snow White. Although dressed alike, they were individuals. The same is true of ordinary scientists in spite of their uniformly white lab coat. Individuality means diversity and there is indeed a great diversity among scientists even when it comes to their scientific philosophy. That many of them subscribe to a kind of operationist philosophy of science should seem to be no surprise. Almost all scientists remain all their lives within their specializations and these become increasingly narrow. Scientists by and large are not so much interested in spotting problems as finding solutions to very obvious and very specific problems, most of the time very practical, technical problems. In other words, since they must produce solutions which work in concrete and immediate contexts they want tools, including conceptual tools, which provide immediate solutions. Hence the kind of work they do makes them subscribe to operationism. The same kind of work makes them form specialized associations, it keeps them within the confines of their own specialized trade, it makes them develop a special jargon, which is particularly suited to the kind of scientific operations which they perform. The kind of operations in question is best compared with the work of a surveyor. He may perhaps ask a question or two about the geological history of the land to make sure that his instruments will be placed on a stable spot. He will not, however, be interested in relating the stretch of land he surveys to the features of the whole continent, let alone to the whole surface of the earth, and much less to the position and distribution of stars and galaxies.

Quite different is the work of the great creative scientists. They, too, have their specialties. Faraday's field was electrochemistry. By training and trade Helmholtz was more a physiologist than a physicist. Much of Planck's work was in thermodynamics. Happily for Einstein, Planck soon saved him from the distractions of a physics teacher by bringing him to the Berlin Academy where Einstein had the sole duty to think in the broadest possible terms about physics and the physical world. Thinking in such terms about physics and the physical world implies a deep concern for the problems of epistemology and for the rationality of the universe. It should be some source of satisfaction that there is a fairly general agreement on this point among philosophers of science. This is true at least in the sense that today it is rather unsafe to make a Copernicus, a Galileo, a Newton, and an Einstein appear as positivists devoted to the "economic" correlation of sensory data.

Contrary to the operationist and positivist clichés about the great creators of exact science, their main concern was a vision of the whole cosmos, a vision steeped in the belief that the whole world was a unity kept together by rational laws. An equally important feature of that vision was that those rational laws could not be simply derived in a Platonic or *a priori* fashion from the preferences of the mind. Plato and his circles were certainly admired by Copernicus, but he was not merely a Platonist. He was a Christian Platonist and this made a world of difference. As a Christian, Copernicus firmly believed that the world was not a self-explaining entity. His Christian faith told him that the ultimate explanation of the world could only be found in the wisdom and will of the Creator. From the wisdom of the Creator it followed that the world had to be fully rational. The will of the Creator implied that the specific pattern of rationality embodied in the world was a choice which man, himself a creature, could not dictate to the Creator. Consequently, man's Platonic or *a priori* speculations had to be shown to agree in all minute details with the data of experience. Those data were of man's gathering but not of his making. While this was very clear to Copernicus, the Christian Platonist, it was almost completely missed by Plato, or by any other purely pagan Platonist.

All these points can easily be found in substance at least in the

Preface and First Book of Copernicus' great work. They stand out very clearly in the writings of Kepler, Galileo, and Newton. Although very clear they are all too often ignored or slurred over in the present-day literature on the sixteenth- and seventeenth-century scientific revolution. It is, therefore, not enough to say that Copernicus was "un bon catholique"[19] and leave it at that. Kepler may have been a sleepwalker, but certainly not when it came to his belief in the Creator. The effort to turn Galileo into a herald of agnosticism[20] entails a systematic oversight of many of his statements. That Newton's faith in God needs Freud for an interpreter,[21] or that the same faith is akin to Newton's preoccupation with alchemy are claims that do violence to very plain texts.

All these and similar efforts seem to be based on the assumption that Christian faith in the Creator has little if any rationality to it. Such an assumption is not so much a reflection on the rationality of that faith as a reflection on the state of mind of those who try to create credibility for it. Psychologically, this state of mind is easily understandable. Few things would seem more difficult than for an agnostic to grasp what it really means to hold and to hold firmly that the whole realm of existence, including one's own very existence, borders on the realm of nothing and if there is existence as we know it, it is only so because of the exclusive power of the One who alone can create. This psychological difficulty is not something to be quarrelled with or to be criticized lightly. Such a quarrel and criticism would be sheer arrogance for which no license can be had in one's faith in the Creator.

Things are somewhat different when it comes to scholarship. Here Christians and agnostics are much in the same boat. The risk run by a Christian historian of Buddhism is similar to the risk run by an agnostic who writes the history of Christian philosophy or portrays the mental physiognomy of a great scientist who is also a firm believer. When for instance Pierre Duhem's devout faith is written off as religious extremism,[22] one cannot help suspecting that a serious narrowing of horizons is at play. Although a very fine physicist, a very incisive philosopher of science, and possibly the most creative historian of science so far, Duhem is a minor figure compared with Copernicus and Galileo. Thus a Duhem can be turned into a practical non-entity. Obviously, the same policy cannot be risked in connection with Copernicus, Kepler, Galileo, and Newton. The only thing that can be done about them by an agnostic historian who is not watching his agnosticism is to minimize the significance of their Christian faith in the Creator for their scientific thinking and to emphasize out of proportion apparently contrary aspects of their mental physiognomy. Such a tactic is feasible as long as one handles their actual statements in a very selective manner. When the statements of Copernicus and others are taken in their fullness they constitute an impressive evidence of the theological aspects of creative science.

About the scientific creativity of Faraday, Helmholtz, and Maxwell two remarks should suffice. One is that their creativity was in a sense a variation on a theme articulated by Newton, namely, that all forces in nature must ultimately be reducible to mechanical laws. The work on the unity of forces as done by Faraday, Helmholtz, and Maxwell was considered a triumph of mechanism, a triumph more spectacular than original in the creative sense. The other remark concerns the broader perspectives in which Faraday, Helmholtz, and Maxwell did their work. For them, and the proof is their having been practicing Christians, the world was susceptible of a creative investigation because the world was the handiwork of a rational Creator.

The evidence in the case of Planck and Einstein should seem all the more telling because it is immune to a standard objection. The objection is that if one is a believer, especially if one is a practicing Christian, one is walking around with a blindfold on one's eyes and one's reflections will instinctively be formulated within a firmly set conceptual framework. Neither Planck nor Einstein can be suspect in that respect. On all evidence Planck was at most a nominal Christian. Such an assumption is strengthened by Planck's negative remarks on the question of the supernatural and on historical, organized Christianity, or the Church in short. In other words, whatever philosophy of science Planck had, he did not owe it to positive Christianity.

This seems also to be borne out from the fact that as a student Planck grew up in an academic atmosphere which was strongly neo-Kantian in its philosophy departments, and strongly anti-philosophical in its science departments. In the latter half of the nineteenth century, scientists in Germany were still aware of the intensive campaign of Schelling, Hegel, and their successors on behalf of a "true" physics. About the champions of that "true" physics Gauss once wrote to an astronomer friend: "Don't they make your hair stand on end with their statements?"[23] Later Helmholtz summed up the German academic conflict with the words: "The philosophers said that the scientists were stupid and the scientists charged that the philosophers were crazy."[24] Scientists could therefore have only contempt for philosophy and especially for metaphysics, and the only kind which was spoken about was idealistic metaphysics. It was with an eye on that metaphysics that Maxwell spoke of the "den of metaphysicians strewn around with dead bones."[25]

Planck certainly did not wish to take up residence in that smelly den. The most natural thing for him to do would have been to adopt an empiricist philosophy which was rapidly gaining in scientific circles largely because of the writings of Ernst Mach. But Mach found his staunchest opponent in Planck. The conflict between them came to a head on the question of whether physics was about external reality or whether it was merely about one's sensations. About that external reality Planck held that it was fully ordered and that such an order existed independently of the thinking mind. Furthermore, Planck perceived that holding these propositions was a metaphysical stance in the sense that it demanded far more on the part of the intellect than a registration and economical organization of scientific data. The additional demand consisted in going beyond physics, or engaging in meta-physics. He never endorsed metaphysics as such, but his whole philosophical conviction was deeply steeped in it.

It was clearly an endorsement of metaphysics when Planck argued that it was legitimate and necessary in physics to assume a deeper layer of physical reality to explain spectral lines, black-body radiation, and the like. The layer below was the realm of atoms. Planck also perceived that when a physicist endorses a layer of physical reality lying beneath what is actually experienced, he is engaged in an act of faith. Planck's repeated statements about the role in science of this spirit of belief are too well known to be quoted here. Somewhat less known are the details of his conflict with Mach. Mach, the radical empiricist, could have no use for metaphysics and for metaphysical belief. Accepting such a belief and going along with the trend Planck represented seemed to him tantamount to joining a church.[26] His remark was uncannily expressive and perceptive. Mach was a good enough philosopher to see what could be the long-range implication of endorsing certain philosophical tenets about the physical world.

Mach was no part of the debate that was to rage about Planck's creative act in science, the quantization of energy. Long before that debate broke into the open in the late 1920s, Planck fully realized that the quantization of energy might suggest a basic disorder in nature. Such a possibility pained him a great deal. As late as 1910, a full ten years after he solved the black-body radiation by postulating discrete or quantized energy emission from atomic oscillators, Planck was still trying to find fault with his derivation of quanta. What seemed to be at stake immediately was the question of mechanical causality in nature. Planck insisted that the fall of mechanical causality cannot mean the absence of ontological rationality in nature. He asserted that rationality with the kind of commitment which is usually ascribed to strong faith. It was with such faith that he opposed the Copenhagen interpretation of science, nature and knowledge. By the time he died in 1947 he had practically only one ally, Einstein. Not a poor ally, if witnesses are weighed and not merely numbered.

It cannot be emphasized enough that Einstein's reluctance to accept the Copenhagen interpretation went far beyond matters of technicality. For him the question at issue was not merely whether (or not) in quantum mechanics a theory of hidden variables could be constructed. He tried it and failed and so did others. For all that he stuck to his philosophical and scientific belief that the cosmos was the embodiment of full rationality, especially in its deepest layer. While his opponents could press him with their famous thought experiments, he kept reminding them of some deeper philosophical issues about the kind of physical reality which is assumed by science. His best known opponent was Bohr, a thinker sensitive enough to see the weight of Einstein's disagreement. Once, while working on the history of his epistemological debate with Einstein, Bohr was found gazing out of the window muttering: "Einstein . . . Einstein."[27]

The story should not distract from an issue which is ultimately philosophical. But the issue at stake here is the historical record. For a historian of science Einstein's philosophical position is a fact, hardly a negligible fact in view of his towering stature as a creative scientist. If a historian decides to evaluate Einstein's philosophical position, he should at least realize that in that case the historian is no longer a historian but a philosopher. He would certainly be mistaken if he tried to do the work of a philosopher with the tools of a historian. A historian still remains within his bounds if he wants to know how Einstein came to take up a position in epistemology which widely separated him from empiricism, sensationism, positivism, and related trends. After all, he started out as an empiricist, an admirer of Hume. Then as a young physicist he thought that he was working according to the prescriptions of Mach. But Mach soon began to suspect that Einstein's reasoning implied views diametrically opposite to his. The matter became crystal clear to Mach after Einstein's work on General Relativity became public. Mach's last writing, the preface to the second edition of his history of optics, was a broadside against Einstein, the philosopher-scientist.

The broadside had one good effect. It shook the scales from Einstein's eyes. He began to see the true physiognomy of the philosophy implied in his creativity in physics. The theory he created, General Relativity, implied that the notion of the world, the universe as a whole, was a valid notion. This put him apart from idealists and empiricists by the same stroke. For Kant and for his followers the notion of the world as a whole was an illegitimate product of the urges of the intellect. Empiricists from Hume on left no stone unturned to discredit the very same notion. They might have done so out of disinterested scholarship, but other motivations might have been at play as well. John Stuart Mill, a prominent empiricist and a candid autobiographer, supports this suspicion. In his autobiography he tells why he wrote his chief work, *The System of Logic,* in which he upheld the possibility that the universe might consist of rational and irrational parts.[28] In the notion of a fully rational universe Mill saw the vindication of metaphysics, and according to the *Autobiography* he deemed it to be of paramount importance to expel metaphysics from what he called its stronghold, namely, its appeal to mathematics and physical science.[29] Only then could society live free of aberration in morals, politics, and religion.

Whether Mill was right or wrong in speaking of metaphysics as being equivalent to idealism and in attributing to such metaphysics serious ties with science, is beside the point. Mill's empiricism could not be reconciled with non-idealist metaphysics, nor with the theism it supported. For that theism the heavens could declare the glory of God only as long as they were ordered throughout. It was hardly an accident that Mill, who advocated the possibility of irrational sectors in the universe, ended up recognizing two gods, both partial, one good and one evil.[30] Undoubtedly, nothing could so badly discredit the glory of one God than cutting the universe into parts of which some were rational and some irrational. While this procedure is compatible with certain philosophies, it is wholly alien to the philosophical framework of creative science as found in the thinking of all great creators of science.

This philosophical framework has been endorsed by Einstein ever more explicitly as he continued reflecting on what he had done in science. The evidence constitutes an extraordinary travelogue of a uniquely creative scientific mind.[31] The record of that travelogue shows Einstein's awareness of two points emphasized in this paper. One is the difference between ordinary and creative science. As Einstein wrote on January 1, 1951, to Maurice Solovine, a long-standing friend: "I have never found a better expression than the expression 'religious' for this trust in the rational nature of reality and of its peculiar accessibility to the human mind. Where this trust is lacking science degenerates into an uninspired procedure. Let the devil care if the priests make capital out of this. There is no remedy for that."[32]

The other point, the theological relevance of creative science, can easily be spotted in another letter of Einstein by anyone familiar with the handling of the traditional proofs of the existence of God in modern philosophy since the time of Hume and Kant. As Einstein wrote on March 30, 1952, again to Solovine: "You find it surprising that I think of the comprehensibility of the world (in so far as we are entitled to speak of such world) as a miracle or an eternal mystery. But surely, *a priori,* one should expect the world to be chaotic, not to be grasped by thought in any way. One might (indeed one *should*) expect that the world should evidence itself as lawful only so far as we grasp it in an orderly fashion. This would be a sort of order like the alphabetical order of words of a language. On the other hand, the kind of order created, for example, by Newton's gravitational theory is of a very different character. Even if the axioms of the theory are posited by man, the success of such a procedure supposes in the objective world a high degree of order which we are in no way entitled to expect *a priori.* Therein lies the 'miracle' which becomes more and more evident as our knowledge develops." To this Einstein added the even more revealing passage: "And here is the weak point of positivists and of professional atheists, who feel happy because they think that they have not only pre-empted the world of the divine, but also of the miraculous. Curiously, we have to be resigned to recognizing the 'miracle' without having any legitimate way of getting any further. I have to add the last point explicitly, lest you think that, weakened by age, I have fallen into the hands of priests."[33] Einstein was then seventy-three.

As everybody knows, Einstein did not fall into the hands of priests but in that eagerness of his to assure Solovine on that score there seems to be more than his well-known bent for fun. In fact his remark is as serious as a cartridge that can explode at the slightest touch. Those who would be the first to touch off that explosion would not be from the ranks of priests or theologians. Rather, it seems to be a safe expectation that had Comte, Mach, and Carnap had the opportunity to read these words of Einstein, they would have exploded. They would rise as a man and ask with a touch of indignation in their voice: "Has it not been proved in Hume's *Dialogue,* in Kant's *Critique,* and in Mill's *System of Logic* that the notion of the universe is not a valid notion? And has it not been proved that anyone who accepts the notion of the universe as fully ordered, has no escape from admitting God as well, and the soul thrown in for good measure?" As a man they would tell Einstein: "It is indeed strange that you fail to see that once you have gone as far as the universe in the way you did, you have no right to say that there is no legitimate way to go any farther."

Although Einstein did not fall into the hands of priests, his creativity in science put him in a philosophical position about which he recognized that it was uncomfortably close to those old, often abused, and almost invariably dismissed proofs of the existence of God. Such is one of the theological aspects of creative science. Another of those aspects can be spotted in the repeated stillbirths of science in ancient cultures. Those stillbirths constitute a monumental pattern in cultural history which should loom even more so in an age of science like ours. Yet in this age of morbid preoccupation with "cultural patterns" this monumental pattern received but slight attention even by historians of science. The reason for this curious neglect is not difficult to identify. It derives from the unwillingness of our age to accept religion except as a cultural pattern. This relativistic approach to religion may serve the disbelief of our age, but it also acts as a blindfold. While the same approach can free a scholar from the perspective of perennial truth in his comparative studies of religion, it will prevent him from facing up to the interaction between science and religion in their historical completeness and true creativity. Obviously, since there is only one viable science, no scholar with indifference, let alone with covert hostility towards Christianity, will consider at length the failure of science in ancient cultures all of which were steeped in paganism. In contrast with these stillbirths of science, its only viable birth took place within a distinctly Christian cultural matrix, hardly cheerful news for the kind of scholar described above.

This difference between repeated stillbirths and one viable birth is a difference between two distinctly theological world views. One is invariably anchored in the idea of eternal recurrence, the other is the fruit of belief in a once-and-for-all creation.[34] For the historian of ideas, or of science in particular, there is another fruitful field to document the theological aspects of creative science. The field is the history of the classic proofs of the existence of God viewed in relation to the development of science.[35] That history shows that all attacks on these proofs when unfolded in their full implication became attacks on the epistemology and world view which proved themselves to be essential ingredients in truly creative science. The correlation is certainly a historical fact. Its detailed account is therefore a proper task for the historian of science not unmindful of the ever present philosophical presuppositions in the methodology of creative science. That methodology also lends itself to a speculative approach which can indeed show that science at its creative best can be a genuine help for the theologian if he is to achieve in this age of science a proper handling of divine truths which are always revealed through human words.[36]

[1]See his letter of March 26, 1598, to Herwart von Hohenburg, in *Johannes Kepler Gesammelte Werke,* vol. XIII, *Briefe 1590-99,* edited by W. von Dyck and M. Caspar (Munich, 1938), p. 193.

[2]*The Herschel Chronicle: The Life Story of William Herschel and his Sister Caroline Herschel,* edited by Constance A. Lubbock (New York, 1933), p. 310.

[3]This is, of course, the broader implication of what Einstein states in his "Autobiographical Note": "Through the reading of popular scientific books I soon reached the conviction that much in the stories of the Bible could not

be true." See *Albert Einstein: Philosopher-Scientist*, edited by P. A. Schilpp (Evanston, 1949), p. 5.

[4]While science is certainly much broader than physics, to speak of science as if it were but physics has some methodological justification. Conceptually, physics is, in a sense, the foundation of all other natural sciences; indeed, most cultivators of the various life-sciences aim at reducing their subject matter to that of physics. The various aspects of that reductionist trend were treated extensively in my *The Relevance of Physics* (Chicago, 1966).

[5]W. Heisenberg, *Physics and Beyond: Encounters and Conversations*, translated from the German by A. J. Pomerans (New York, 1971), p. 215.

[6]For a recent expression of this conviction, see J. M. Ziman, *Public Knowledge: An Essay concerning the Social Dimension of Science* (Cambridge, 1968), pp. 39 and 144.

[7]C. P. Snow, *The Two Cultures and a Second Look* (Cambridge, 1964), p. 10.

[8]See note for the second printing, New York, 1952, p. [vii].

[9]H. Bondi, *Cosmology* (2nd ed.; Cambridge, 1960), p. 144.

[10]See especially C. W. Taylor and F. Barron, eds., *Scientific Creativity: Its Recognition and Development* (New York, 1960), and A. Koestler, *The Act of Creation* (New York, 1964).

[11]For an emphasis on the historical aspect, see R. Taton, *Reason and Chance in Scientific Discovery*, translated by A. J. Pomerans (New York, 1962). The conceptual aspect is stressed in W. I. B. Beveridge, *The Art of Scientific Investigation* (rev. ed., New York, 1957); in N. R. Hanson, *Patterns of Discovery* (Cambridge, 1958); and in R. J. Blackwell, *Discovery in the Physical Sciences* (Notre Dame, In., 1969).

[12]Thus, for instance, L. S. Feuer, *Einstein and the Generations of Science* (New York, 1973). Feuer is also the author of *The Scientific Intellectual: The Psychological and Sociological Origins of Modern Science* (New York, 1963).

[13]The works of this kind that created the greatest following are *Personal Knowledge*, M. Polanyi (London, 1958), and *The Structure of Scientific Revolutions*, T. S. Kuhn (2nd ed., Chicago, 1970).

[14]This literature is largely the outgrowth of the studies on Galileo by A. Koyré whom R. S. Westfall aptly called the "dean of historians of the scientific revolution." See R. S. Westfall, "Newton and the Fudge Factor," *Science* 179 (1973), p. 751.

[15]M. Planck, *The Philosophy of Physics*, translated by W. A. Johnston (New York, 1936), p. 97. In the same context (p. 96) Planck referred to Mach as one of those who claimed to themselves an authority "simply above argument."

[16]See the first three chapters in my *The Relevance of Physics*, quoted in note 4 above.

[17]As reported by M. Delambre in his "Notice sur la vie et les ouvrages de M. le comte J. L. Lagrange" in *Oeuvres de Lagrange* (Paris, 1867-92), vol. I, p. xx.

[18]The English translation itself went through six editions of which the last corresponded to the ninth German edition. Mach supported his view with a reference to Gauss who noted that no essentially new principle can be established in mechanics. *The Science of Mechanics: A Critical and Historical Account of its Development,* by Ernst Mach. Translated by Thomas J. McCormack (6th ed., La Salle, Ill., 1960), p. 441.

[19]A. Koyré, *La révolution astronomique* (Paris, 1961), p. 19. However, two pages later Koyré describes Copernicus as a humanist "dans le meilleur sens du terme," a sense which Koyré does not explain.

[20]L. Geymonat, *Galileo Galilei: A Biography and Inquiry into his Philosophy of Science,* translated by Stillman Drake (New York, 1965).

[21]Frank E. Manuel, *A Portrait of Isaac Newton* (Cambridge, 1968).

[22]D. G. Miller, "Pierre Duhem," *Physics Today,* 19 (1966), p. 53.

[23]*Carl Friedrich Gauss Werke* (Göttingen, 1870-1933), vol. XII, p. 62.

[24]"On the Relation of Natural Science to General Science" (1862), in *Popular Lectures on Scientific Subjects,* translated by E. Atkinson (New York, 1873), pp. 7-8.

[25]Address to the Mathematical and Physical Section of the British Association" (1870), in *The Scientific Papers of James Clerk Maxwell,* edited by W. D. Niven (Cambridge, 1890), vol. II, p. 216.

[26]See P. Carus, "Professor Mach and his Work," *Monist* 21 (1911), p. 33.

[27]As reported by an eye witness, A. Pais, in S. Rozental, ed., *Niels Bohr* (Amsterdam, 1967), p. 225.

[28]*The System of Logic, Ratiocinative and Inductive,* Book III, chapter xxi §1. The chapter in question is "Of the Evidence of the Law of Universal Causation."

[29]*Autobiography of John Stuart Mill* with a Preface by John Jacob Coss (New York, 1924), p. 158. This edition is the first to contain the integral text of the manuscript.

[30]See especially Part II, "Attributes," in *Theism,* which in turn is a part of Mill's *Nature, the Utility of Religion, Theism, Being Three Essays on Religion* published a year after his death.

[31]G. Holton, "Mach, Einstein, and the Search for Reality," *Daedalus,* Spring 1968, pp. 636-73. Holton's article begins with a characterization of Einstein's mental development as a "pilgrimage from a philosophy of science, in which sensationism and empiricism were at the center, to one in which the basis was a rational realism." The following quotations are not in Holton's article.

[32]Albert Einstein, *Lettres à Maurice Solovine,* reproduits en facsimilé et traduites en français (Paris, 1956), pp. 102-3.

[33]Ibid., pp. 114-15.

[34]This contrast forms the basis of my *Science and Creation: From Eternal Cycles to an Oscillating Universe* (Edinburgh, 1974). See also my article, "God and Creation: A Biblical-Scientific Reflection," *Theology Today,* 30 (1973), pp. 111-20.

[35]This is the subject of my Gifford Lectures given at the University of Edinburgh in 1975 and 1976 under the title, *The Road of Science and the Ways to God.*

[36]It is no small satisfaction to me as a historian of science that the pioneering work in this respect is done by the very same theologian, the Right Rev. Dr. Thomas F. Torrance, to whom this essay is dedicated. See especially his "The Integration of Form in Natural and in Theological Science," in *Science, Medicine and Man* (London, 1973), vol. I, pp. 143-72.

10

The University and the Universe

I am convinced I can speak in the name of all present, benefi-
ciaries of a provident Alma Mater, which in her generosity imposes
upon us only one obligation, the pleasant duty of exchanging ideas
across the universal field of knowledge. Ours is a state of sweet do-
nothing, a sort of *dolce far niente*; no courses, no exams, no papers
to correct. Our sole business is learning for learning's sake. Two
famous educators, were they here with us, would point out that
such a gathering has an uncanny likeness to what they held to be
the ideal of a university. John Henry Newman, as you all know,
held high the view that the university is an institution which "merely brought a number of young men together for three or four years
and then sent them away." He contrasted it with that organization
which "gave its degrees to any person who passed an examination
in a wide range of subjects." The contrast was also between an
institution which to all appearances "did nothing" except insist on
residence of students and tutors, and an organization "which

Paper read at the Faculty Retreat of Pepperdine University, October 9-10,
1981, and reprinted from its Proceedings, *Freedom, Order and the University* (Malibu, California: Pepperdine University Press, 1982), pp. 43-68.

exacted of its members an acquaintance with every science under the sun."[1] The other educator, Mark Hopkins, president of Williams College and a contemporary of Newman, went about his business with an informality and naturalness which evoked the best in the American character and prompted James Garfield, a Williams alumnus, to his famed remark: "The university is a student on one end of a pine log and Mark Hopkins on the other."[2]

In this graphic picture of a student and a teacher doing in all appearance nothing but shooting the breeze, Newman would, in spite of all his love of the ancient formalities of Oxford, recognize his ideal of the university. He would be horrified at seeing the almost complete triumph of the concept of the university "as a factory of knowledge," a concept urged by T. H. Huxley,[3] a contemporary of both Newman and Hopkins. In such a factory he would easily recognize that modern version of universities which produce "numerates" instead of "literates"[4] and in which information passes from the mouth of the teacher to the ears of the student without having passed through the minds of either of them. It should be easy to guess Newman's reaction to universities in which a dissertation on time and motion comparison of four methods of dishwashing qualifies its author to the dignity of master of arts.[5] Churchill, who once spoke of "some chicken, some neck," might now say, "some masters, some arts." Newman would undoubtedly relish the concluding part of the remark according to which "Harvard offers education *à la carte,* Yale a substantial *table d'hôte,* Columbia a quick lunch, and Princeton a picnic."[6] A picnic has something leisurely in it, almost invariably it includes the presence of one or more logs and above all the atmosphere in which thoughts flow unencumbered. Newman would, of course, shake his head in disbelief on hearing that some universities became, to quote a Chinese student visiting in the United States, "athletic institutions in which a few classes are held for the feeble-minded."[7] One can easily imagine Newman's revulsion to universities which turned into a grim game of musical chairs: students claiming to themselves the role of teaching, teachers wanting to administer, and the administration lamely footing the bill for the self-destructive tragicomedy. Newman would see tragedy loom large in a university which has for its switchboard rule that an outside or inside caller must give the number of a student, not his or her name. The university was for Newman an Alma Mater, "knowing her children one by one, not a foundry, or a mint, or a treadmill,''[8] or a computer printout, he would add, were he alive today. Great lover of books though he was, he would be dismayed by the phrase, old yet applicable in some places today, that a "true University of these days is a Collection of Books."[9] He would find revolting the spectre of universities eager 'to make young men as unlike their fathers as possible.''[10] Mindful of the old truth that the new morality is the old immorality, he would not be overly shocked on finding universities that "are fit for nothing but to debauch the principles of young men [and women].''[11] But nothing would shock him so much as the abdication of the search for meaning when voiced by spokesmen of illustrious universities, whether its presidents or eminent professors, especially when they do it in the name of knowledge.

Certainly, Newman was not an enemy of knowledge. Apart from his universally acknowledged scholarship, his idea of the university most explicitly contained a search for all knowledge, for knowledge in its universality. But precisely because he believed in the universality of knowledge, he held that knowledge was pursued for the sake of meaning, and above all for the sake of that most universal meaning which is God, the ultimate in intelligibility and being. This is why he made it the core of his idea of university that a university retain its *raison d'être* only if philosophical or natural theology was the basis of its program of instruction.[12] Newman also knew that search for meaning was best done when the mind was free to reflect on and share in all kinds of knowledge, not only free in a political sense, but also free of psychological and social coercion. This is why he esteemed so highly the unstructured encounters between students and tutors, this is why he would have recognized his idea of a university in a pine log reserved for one student and one teacher. It is in such encounters that one asks the questions: What is the point of knowing or is knowledge ultimately pointless?

Most present-day universities, and especially the most prestigious ones, have, to all appearances, given up the search for meaning. Illustrations of this are the sundry utterances of especially two groups of scholars in great modern universities, cosmologists and biologists. They tower above their colleagues not so much because they have at their disposal dazzling equipment and control budgets vastly superior to those allotted other departments. Their superiority does not even rest on the all-inclusive and fundamental character of their empirical subject-matter: galaxies and genes. Their commanding role rather lies in the fact that they have somehow expropriated to themselves an authoritative role to think about universe and man. Of course, modern cosmologists (or fundamental particle physicists—their fields more and more overlap) and modern geneticists must think more than ever. From Einstein to Hawking, modern cosmology has been above all a feat of incisive thinking which almost invariably was a step or two ahead of observation. Genes had for long been a conceptual postulate before they were actually isolated, analyzed and manipulated. Even today— when gene splicing is turning into a big industry and when leading universities are eager to secure millions from genetic discoveries made in their laboratories—it is well to recall that not so old story of the double helix. The story was more the story of a race among minds than of a race among laboratories.

Scientific thinking certainly deserves the highest admiration, but here too, as elsewhere, admiration can be easily misplaced. What is all too often forgotten about scientific thinking is that it is primarily thinking and only secondarily scientific. Even the most trivial scientific statement is steeped in philosophy, and to be blunt, steeped in plain metaphysics. The amount of metaphysics in scientific statements increases in the measure in which these statements become more inclusive. At the point where these statements become valuational, their tie to truth will not be scientific at all but thoroughly philosophical or metaphysical. First-rate scientists should therefore provoke not admiration but consternation when after frowning at length on philosophy they wax utterly philosophical. A case in point is the concluding remark in S. Weinberg's masterly popularization of modern cosmology, *The First Three Minutes*: "The more the universe seems comprehensible, the more it also seems pointless."[13] A more concise and telling illustration of the ultimate tragicomedy which is the hallowed atmosphere of great modern universities, could hardly be found. Unlike some cosmologists, who today celebrate the lack of comprehension, biologists have for more than a hundred years revelled in equally crude philosophical non-sequiturs. In particular I mean their purposeful debunking of purpose. They not only get away with it, but earn interminable applause for the farce.

One example, and a fairly recent one, the reception of Prof. E. O. Wilson's book, *On Human Nature,* should suffice. Although it comes to a close with the admission that "the mythology of scien-

tific materialism" is both liberating and enslaving through its
ultimate offering which is "blind hope," the book earned accolades
for its philosophical merits. A book which proclaims in the same
breath both liberation and enslavement should have appeared suspect prima facie, but not in this age of modal logic. Only the old
fashioned disjunctive logic entitles one to ask that if man, as Prof.
Wilson insisted, was merely a system to secure the survival of viable
genes,[14] where was his liberation? Is it not a cruel joke to cheer up a
prisoner, never to be released from jail, by telling him that he
should feel liberated through being informed about his unchange-
able imprisonment? Again, was it not rudely farcical on the part of
Prof. Wilson, praised by another reviewer as "a thoughtful scientist,"[15] to claim that *On Human Nature* was "the third book in a
trilogy that unfolded without my being consciously aware of the
logical sequence until it was nearly finished?"[16] No readable book,
not even the opium-eater Coleridge's *Wanderings of Cain* and
Kubla Khan, was ever written without its author having a conscious advance view of its plan and message.[17] What makes *On
Human Nature* not only a very readable, but also a very coherent
and instructive book is, in fact, the very force of premises to which
Prof. Wilson is consciously captive throughout the book.

That the last pages of the book should ring of despair, should
seem a foregone conclusion to anyone ready to pause over the
startling motto of the book, a quotation from David Hume. This
was perhaps not clearly foreseen by Prof. Wilson. But it is testi-
mony to his greatness as a thinker (his greatness as a scientist does
not need my praises) that he stuck to his premises through thick
and thin, even at the price of ending with the patently contradictory message of simultaneous liberation and enslavement.
Perhaps he could not simply extricate himself from the hold of
system-making, the lure of which is much stronger than generally
suspected. At any rate, Prof. Wilson's contradictory message and
his claim that purpose can nowhere be seen, should seem farcical
even on a cursory look. The fact is, however, that most academic
faces turn rather sour when reminded of Whitehead's famous
remark, made half a century ago, that scientists who devote themselves to the purpose of proving that there is no purpose, constitute
an interesting subject for study.[18] The tragicomedy is indeed complete. In modern universities, all too often equipped with plush
theaters, faculty weep over their performance when they should

laugh, and smile when they should hang their heads in shame.

The immediate cause of their shame should be their playing a facile game with philosophy. They do not seem to have taken for more than an artful fishing for compliments the dictum of Einstein, a most eminent among scientists, that "the man of science is a poor philosopher."[19] Quite possibly Einstein himself did not realize what a poor philosopher he was. He could never articulate that realism and its meaningfulness to which he was driven by his scientific creativity.[20] To the end he made relapses either into Kantian idealism or Machist sensationism, the two philosophies which he imbibed as a youth, and both of which he disowned once he perceived the true meaning of his work in physics. As all too often in the past, in our century too, it remained to others, usually philosophers, to articulate the deeds and words of leading scientists, and to unveil, if necessary, the tragicomedy lurking behind some of their utterances. Such a role on the part of the philosophers should not seem to be surprising as most of the time scientists speak of two heavily philosophical subjects, man and the universe. Both are vast topics even for a series of lectures, and are also very different topics. Yet they have one very important trait in common: both man and universe are invisible to physical eyes. That we cannot take a physical look at the universe because we, being a part of it, cannot get outside it, should be obvious. As to man, I do not wish to rekindle old debates about universals of which man, Socrates, was the chief whipping boy. To those who dismiss that debate as a scholastic play with words, let me recall a story about Claude Bernard, certainly not a scholastic and not even a philosopher by profession. When asked whether the study of life demanded a mechanistic or a vitalistic philosophy, he curtly replied: "I have never seen life,"[21] with physical eyes, of course. The phenomenon of mere organic life lands us deep into metaphysics. And so does the universe.

This last remark should seem particularly appropriate in this year of 1981, the 200th anniversary of the publication of Kant's *Critique of Pure Reason.* If there is a book of which cosmologists should be wary, it is the *Critique.* The same holds true for all scientists, if it is true, and it certainly is, that all science is cosmology.[22] For if the *Critique* is right, cosmologists and all scientists can only be wrong, and should consider themselves to be the victims of a tragic illusion. For a chief aim of the *Critique* is that the notion of cosmos, or universe, is not a valid notion; in fact, to recall Kant's claim, the notion of the universe is merely the bastard product of the metaphysical cravings of the intellect.[23] Actually, it is Kant's procedure to validate this claim that should seem both dastardly and clever at the same time. The cleverness relates to the fact that he first took on the universe and only afterwards man. The cosmological antinomies precede the anthropological ones. Once the validity of the cosmos is undermined, it is far easier to lock man within himself and secure for him thereby an absolute autonomy, the ultimate aim of the *Critique,* though a much earlier aim of its author.[24] The dastardliness of the antinomies as articulated by Kant is glaring. He shifted grounds from idealism to empiricism, choosing now the one, now the other, to suit his immediate strategy which aimed at enveloping both universe and man in a fog of uncertainty. As to the universe, idealism provided his argument that the universe could not be demonstrably finite, and he fell back on empiricism to support the opposite, namely, the impossibility of proving the infinity of the universe. A universe which thus appeared to be without contours was for Kant an entity which man tried in vain to grasp, let alone to use as a stepping stone to the ultimate of being and intelligibility, usually referred to as God.

Were Kant alive today, he would be tormented by the fact that the very same science, which he took for his stronghold, though he knew very little about it,[25] achieved rigorous ways to deal with the totality of consistently interacting things, the universe. In this century of science explosion, we are often overawed by our ability to explode the atom and far less impressed by a much more explosive scientific achievement, the formulation for the first time in history, of a genuinely scientific cosmology. The success in that latter respect is far from being complete. Things interact in more than one way, not only gravitationally but also electromagetically. There are in addition the nuclear forces binding the nucleus, and the force binding the quarks, the constituent part of nucleons (protons and neutrons), a force which is the subject of the so-called chromodynamics. The formulation of a Unified Theory in which all these forces, and perhaps some others, will be seen as the manifestation of one single force or interaction, is the great aim of leading scientists today. A chief reason of their confidence relates to Einstein's General Theory of Relativity. In General Relativity scientific cosmology found its true birth insofar as the theory gives a paradox-free account of all material bodies which consistently interact with one another gravitationally.

The magnitude of Einstein's achievement can only be perceived when set against a rather dark background. It consists of the insensitivity which scientists displayed throughout the eighteenth and nineteenth centuries toward the gravitational and optical paradoxes of an infinite Euclidean homogenous universe of stars. The paradoxes were sufficiently stringent to suggest the impossibility of such a universe. To my knowledge, only two scientists, Zöllner in Leipzig and Clifford in London, did recognize in full, around 1870, that a lecture of Riemann, given in 1854, provided the possibility of a paradox-free, that is, contradiction-free cosmology.[26] The name of Riemann should evoke non-Euclidean geometry and a four-dimensional space-time manifold. Indeed, Einstein's General Relativity, or at least its cosmological chapter, was not without some anticipations, none of which made a real stir.[27] The idea of an infinite Euclidean universe so strongly dominated the thinking of scientists around 1900 as to make them adopt a schizophrenic state of mind. They took the Milky Way for an all-inclusive entity. While they believed that stars and galaxies stretched to infinity, they conveniently wrote off everything beyond the Milky Way as irrelevant for science.

Beneath this infatuation with Euclidean infinity there lay considerations that were inspired by Plotinus' pantheism. This is not to suggest that Pseudo-Dionysius, a still-elusive sixth-century author, wanted to graft pantheism on Christian thought as he propagated the notion of God's infinity in a distinctly emanationist or Plotinist sense.[28] He took God's infinity in a positive sense, in a distinct departure from Christian tradition, for which the term is negative, a tradition which on this point too was emphatically voiced by Thomas Aquinas. During the centuries of decaying Scholasticism less and less attention was given to the dangers implied in the notion of infinity. Even in the case of Cusanus, piety and scientific acumen were not enough to keep those dangers in focus. Distrustful of philosophy, Cusanus declared under distinctly Plotinist influences that the creation must be similar to the Creator. With some strict reservations the declaration can still be Christian; with no such reservations it opens the floodgates first to pantheism and then to strict materialism in which the infinite universe unashamedly plays the role of God. Indeed, a Giordano

Bruno, a champion of pantheism, is Cusanus without Christian faith and piety. The parallels among the dicta of the two are legion, or rather Bruno plagiarized Cusanus to an astonishing degree. One illustration of these parallels should suffice. Cusanus did his best to eliminate the distinction between planets and stars. With great vehemence Bruno did the same. Indeed the chief characteristic of the universe, both in Cusanus' and in Bruno's version is that it tends to lose its clear distinct specific features.[29] The presence of Plotinist theology in Newton's cosmology, a further major step in the story, is best seen in Newton's insistence on the infinity of space as sensorium of an infinite God. Kant, that is the young Kant, author of a cosmological work, merely echoed an all too common theme of his time, when he spiritedly argued that only an infinite creation is worthy of an infinite god.[30] Before long, Kant needed no God, or even a universe, for that matter.

A chief difficulty which plagued Newtonian physics was the ether, a material entity which filled infinite space. The ether, as is well known, could not be given paradox-free properties. At this point nothing would be more tempting than to jump to Einstein. Is not he, in all the cliché-ridden accounts, the glorious St. George who beheaded that dragon, the science of the ether? The cliché is certainly effective in drawing attention away from a crucial chapter in that theological story about infinity. In that crucial chapter many nineteenth-century agnostics and materialists would have room, but none so prominently as Herbert Spencer. Spencerian philosophy is an elaborate materialist theology of an infinite universe.[31] The chief characteristic of that theology is an effortless flow of words in which the emergence of the specific from the non-specific, the differentiated from the undifferentiated, is being set forth. Herbert Spencer is the nineteenth-century Giordano Bruno who in a discourse, which contains nothing of Bruno's crudities and which, unlike Bruno's discourse, has all the semblance of science, dissolves the specificities of the universe into an endless sea of non-specificities. Herbert Spencer, it is well to recall, was a foremost champion of the nebular hypothesis, or Kant-Laplace theory, the starting point of which is a nebula which, as far as science was concerned, was nebulosity itself,[32] that is, the absence of all specificity. Reference to Spencer and specificity should remind one of Darwin, whose message was far more than the plea on behalf of the emergence of a species from another. The closeness of the word species to specificity is in itself suggestive of another far deeper issue. It is the story of the instinctive struggle of materialists (Darwin was an avowed one from the start)[33] against cosmic specificity. The cosmological dicta of Marx, Engels, and Lenin, tell the same story.[34]

Chronologically, we are now at the turn of the century when scientists, both believers and non-believers, firmly asserted the infinity of the universe, on spurious theological or counter-theological grounds, and, in one way or another, tried to talk away the specificity of the universe. The reasonings of both groups were distinctly non-scientific.[35] Einstein's main achievement was to put cosmological discourse on truly scientific tracks. Here too misleading clichés abound. The heart of the cosmology which General Relativity made possible, is not whether the total mass of the universe is finite or not, or whether it is expanding or not, or whether its expansion would turn into a contraction or not, let alone whether the universe is 10 or 18 billion years old. The heart of that cosmology is the value, a most specific figure, which it is able to give about the space-time curvature valid for the entire universe. Most likely, that value is a small positive number, standing for the closed spherical net of permissible paths of motion, the new definition of space. But even if that curvature should turn out to be a small negative number, standing for a hyperbolic space, it should still strike one with its specificity. Such a space can be illustrated by a saddle, with no edges, but with extremely well-defined slopes. The only possibility which is certainly excluded is Euclidean infinity whose curvature is 0, an age-old symbol of non-existence.[36]

Whatever has been learned about the cosmos since Einstein published in 1917 his cosmology provided further stunning details about the specificity of the universe in space as well as in time. In Prof. Weinberg's *First Three Minutes* there are some beautiful pages on that specific primordial soup which gave rise to the chemical elements on which all else, including our own specific existence, is based.[37] Specificity is the hallmark of all the avidly pursued cosmological research which pushed the study of the genesis of the universe beyond its baryon stage, described in the *First Three Minutes*. The earlier states investigated more recently are known as the lepton, hadron, quantum, and matter-antimatter states. In those cases we see the same story: the story of one cosmic specificity leading to another, and in staggeringly exact and specific quantitative terms. In the matter-antimatter state, for instance, ordinary matter particles must outnumber antimatter particles in the specifically exact ratio of one part in ten billion to let subsequent physical interactions issue in processes characteristic of our actually observed specific universe.[38] At any stage, the slightest departure from the specificity as postulated would prevent the formation of galaxies and certainly the emergence of man. This is the consideration which made so many cosmologists for the past twenty years speak of the anthropic principle.[39] The principle stands for the nagging suspicion that the universe may indeed have been fashioned for the sake of man. Clearly, cosmologists are, in spite of themselves, in the grip of a meaning which stretches from the universe to man and from man to the universe and beyond.

There are of course brave, never-say-die warriors who try to undercut this most momentous outcome and perspective which are replete with theology. Their efforts are either scientifically self-defeating, or scientifically revealing, or logically impossible. Into the first class belongs the now-defunct steady-state theory, whose proponents tried to save infinity for the universe along the time parameter only to pile infinite matter upon infinite matter since eternity. Into the second class belong the efforts which try to restore perfect homogeneity to the space-time manifold, or rather to the vacuum postulated by quantum field theory. In the context of such efforts one comes across baffling questions about the strangely specific ratios of physical forces and interactions and one finds revealing comments such as: "Some or all of these questions may not have answers. The world may be just the way it is."[40] The third class is made up by theories aimed at showing on an *a priori* basis that the structure and extent of matter of the material world can only be what they are and nothing else. What these theories aim at is not a complete description and account of all known phenomena. This is in itself not impossible. The physical universe ought to be thoroughly ordered and rational, if the findings of science should have any meaning at all. The famed astronomer James Jeans made a notable muddle of philosophy, and even more of science, when he entertained the world with a once famous book, *The Mysterious Universe.*[41] Furthermore, it is not in itself impossible that one should by a lucky stroke of genius hit upon a mathematical formula which could deal not only with all known phenomena but would prove equally successful with all phenomena still to be discovered about the universe. Some theoreticians (the third class in question) dream about much more. They hope to construct a mathematical physics which would be equivalent to showing that the structure or specificity of the universe can only be what it is and nothing else. Such hopes, as I have kept insisting in published writings and in open oral debates for the past fifteen years, though apparently without creating any echo, should be viewed as logically impossible, as long as Gödel's incompleteness theorem is valid.[42] In other words, the specificity of the universe will remain the kind of specificity which keeps reminding any sensitive mind that it is not a necessary but a contingent feature, a specificity which does not have its *raison d'être* in itself, but must depend on a choice external to the universe.

This is the very core of the message of modern cosmology about the universe. That most cosmologists are reluctant or simply unable to spell it out clearly, or take evasive action about it, does not matter. The message is there and is recognized by sensitive minds in private at least. Solovine, a friend of Einstein, shared his concern only with Einstein that Einstein's cosmology evoked the existence of a Creator. Nor did Einstein publish his reply which contained phrases, reassuring on a cursory look but full of strange uneasiness in between the lines: "I have not yet fallen in the hands of priests," Einstein wrote to his friend, and added: "Let the devil care what priests would do with my cosmology."[43]

What priests do with Einstein's cosmology, with his science of the universe, should not be of any concern here. What should be of concern here is what the universities are doing with the universe as revealed by twentieth-century scientific cosmology. For, clearly, unless etymologies are completely misleading, the universe and the university cannot be foreign to one another. A university as an institution was born in the belief (a belief specific to the Middle Ages) that it is meaningful to search for universal knowledge, precisely because there is a universe, that is, a coherent totality of things and minds. Are universities still such institutions, or have they degenerated into places of entertainment where non-science students are initiated into twentieth-century cosmology through courses in which the mythology of extraterrestrial intelligence is presented with all the dazzling glamor of audio-visual techniques as the latest in respectable and reliable science?[44] Telling signs of intellectual degeneracy can also be gathered from the various concerns which aim at bridging the gap between the two cultures, the humanities and the sciences, and above all from the patently sick condition in which the humanities find themselves. Humanists blame as a rule the encroachment of the quantitative scientific method in all fields, even in fields where they have hardly anything applicable. Here humanists certainly have a valid point. Quantitative method, when applied exclusively in the field of values, leads to an erosion of meaning, an erosion which does not seem to worry many scientists secure in their chairs, laboratories, and shore-side bungalows. But others, especially the young who feel all too keenly uncertainties of all kinds, cannot be fooled. It was in all likelihood someone very familiar with the depersonalized atmosphere of modern universities who wrote the poignant lines:

> Lost in concrete canyons
> and captivated by our knowledge of science
> we tend to see God nowhere.[45]

Clearly, the young complained about lack of meaning. The great scouts of the canyons of science did not provide meaning, quantities in themselves never do. What about the humanists? Many of them turned into pseudo-humanists who blindly ape the scientists. They trust scientific data and computer printouts much more than their own minds and remain unaware of the third of the three kinds of lie: plain lie, big lie, and statistics. Other humanists, the minority, still mindful of the mind, are largely demoralized. The result was put concisely in a comment on the Rockefeller report on "The Humanities in American Life": "Humanists have lost their franchise. . . . Humanity goes on without the humanists."[46] If only humanity would go on, but obviously it does not. A mere look at the tensions, ready to rip apart the globe, although sprinkled all over by thousands of universities, should be enough of a proof. In fact, as a recent study showed, of all institutions it is the universities that most easily accommodate professional instigators of tensions.[47] It is all too obvious that humanists themselves have lost, by and large, confidence in meaning and for two reasons: One is that they have lost trust in human nature as something more than a machine. They simply caved in to scientism propagated by shallow scientists and scientifically uninformed philosophers. The second reason is that they have no close contact with that minority among leading scientists who are very sensitive to the limitations of the scientific method. But whether these remarks are right or wrong, the atmosphere of universities is hardly conducive to questions about meaning. Knowledge they certainly give, but the more they give the greater is the hunger for meaning.[48] To those ready to brand these remarks as the futile lamentations of a priest let me quote the Grand Master of the Grand Orient in Paris. In connection with the sweeping changes effected in French universities by the Socialist government (13 of the 27 French universities have already received new rectors—a process which witnessed the replacement of Giscardian technocrats with social engineers) the Grand Master said: "We Freemasons do not accept the transformation of our universities into productivist establishments. The mission of universities is the teaching of the humanities and the sciences."[49]

The obvious failure of universities, the chief teaching institutions, to live up to the goal of the kind of teaching which is a search for meaning, that is, human understanding, is strongly suggestive of a situation which J. S. Mill described in prophetic words:

> When the philosophical minds of the world can no longer believe its religion or can only believe it with modifications amounting to an essential change of its character, a transitional period commences, of weak convictions, paralysed intellects, and growing laxity of principle, which cannot terminate until a renovation has been effected in the basis of their belief, leading to the evolution of some faith, whether religious or merely human, which they can really believe; and when things are in this state, all thinking or writing does not tend to promote such a renovation, is of very little value beyond the moment.[50]

The new religion, or the new meaning, which Mill had in mind, was an outlook steeped in science. It may surprise you but I have no objection to this religion and for a very simple reason. The more genuine success is claimed by science, the more specific the universe will appear. Of course, any aspect of ordinary reality is very specific, specific to the point of being queer. The queer specificity of the real world immediately surrounding us, the everyday world, has a lasting freshness only to very sensitive onlookers and extraordinary minds. But the ordinary mind cannot help being startled when it finds the entire universe described by science in a few bafflingly specific terms. Even the ordinary mind would be awakened to the fact that such a specificity is hardly an exclusive possibility. Once this is realized, Creator, religion, and meaning emerge on the mental horizon and beckon for recognition. Had Chesterton lived to learn about modern cosmology, his always voluble enthusiasm would have overflown. For already in 1905 when he wrote his first major book, *Heretics,* he fully saw that the great disease and heresy of modern culture was its refusal to consider the general, that is, the meaning. Modern culture, as Chesterton put it, was lost in details and in a state of mind in which "everything matters—except everything,"[51] that is, the universe. Modern culture, precisely on account of its professional agnosticism if not plain paganism was in the strait-jacket of a blind dogmatism which prescribed, to quote Chesterton again, that "Man may turn over and explore a million objects, but he must not find that strange object, the universe, for if he does, he will have a religion, and be lost."[52] If such was the case, and Mill, who spoke of cosmology as the stronghold of theists,[53] would certainy have agreed, then it followed, to paraphrase a remark of Chesterton, that once our students were given the universe, they would have religion.[54]

A better formulation of what should be the chief mission of a healthy university today to the sick world of today could hardly be put any better. The university must give the universe so that the world may have religion, or else the world will have a counter-religion, a much worse predicament than all the disasters which true religion may, in spite of itself, promote. Healthy universities would also find in Chesterton's *Heretics* a marvelous diagnosis of the inner logic which sets two leading departments in today's universities at loggerheads: the cosmologists and the biologists. The cosmologists, being in the grip of the specificity of the whole, will instinctively hand down a teaching very different from the one given by the biologists, who are all too often microbiologists. Concerned with the minutest details they can easily grow insensitive to the whole, that is, man and his search for meaning. "Before long," Chesterton prophesied in 1905, "the world will be cloven with a war between the telescopists and the microscopists."[55] The war, or conflict, is all too evident today. While cosmologists all too often are forced to face up to the question of meaning, posed by an unbelievably strange universe, biologists, especially their Darwinian kind, blissfully go on with their narrow strategy which leaves no room for questions about meaning. Typically, it is when they take a quick look at what their cosmologist colleagues are doing that it dawns suddenly on them that the strategy may be too superficial to secure ultimate victory for them. A most illuminating case of this occurs in the very first page of the very first chapter of Prof. Wilson's *On Human Nature*. The case is illuminating in more than one sense. While the illuminating force of cosmology can be sensed between the lines, each of those lines is a sad philosophical muddle:

> If humankind evolved by Darwinian natural selection, [then] genetic chance and environmental necessity, not God, made the species. Deity still can be sought in the origin of the ultimate units of matter, in quarks and electron shells (Hans Küng was right to ask atheists why there is something instead of nothing) but not in the origin of the species. However much we embellish that stark conclusion with metaphor and imagery, it remains the philosophical legacy of the last century of scientific research.[56]

A legacy which is the hybrid compound of chance and necessity is anything but philosophical. Neither Prof. Wilson, nor other Darwinists, and certainly not Prof. Monod or any of the members of the Copenhagen school of quantum mechanics, have ever given a philosophically satisfactory definition of chance.[57] The best definition of chance is still that it is a convenient cover for our ignorance. As to the crediting of Hans Küng with the specific question, why there is something rather than nothing, it is worthy of a scientist who knows theology only through the good services of the Book of the Month Club. Indeed the philosophical and theological poverty of Prof. Wilson's lines could not have been greater, a poverty typical of leading faculty members of modern universities. For if indeed God, the ultimate in intelligibility and being, that is, meaning, can be sought in the origin of quarks, electron shells, 3°K radiation, over-all cosmic space-time curvature, in cosmology in short,[58] then the claim that the evolution of species is a meaningless chance process will have no foundation. Once there is a Creator, there can be no chance in the pseudo-ontological sense which Darwinists attribute to that word. They should above all remind themselves that if all is chance and necessity or both, then any scientific discoverer, and certainly any Nobel Prize winner, must carefully avoid crediting *himself* with any discovery, big and small.

Clear and cogent as this may be, acknowledgment of it will not come forth from most of the leading academics captivated by what they see through their microscopes. Even most of those who look through their telescopes will take refuge either in solipsism,[59] or in some variation of Bertrand Russell's claim that nothing forces man to look beyond the universe.[60] Nothing, of course, except the intellect insofar as the intellect is *non*-mechanical. Such an intellect cannot be forced in a mechanical sense, not even in a sense which is touted to be logical but, like all equations of mechanics, is tautological because logic deals with strict identity relations. The most overrated and at the same time most explosive symbol of mathematical physics is that equation sign which states that nothing happens unless something is already happening. In the absence, or rather precisely because of the absence of mechanical necessity in the process of demonstration, the confrontation of believers and non-believers will go on. Let us hope that, in the best interest of those on both sides, it will go on in that liberal spirit which is not blind to the difference between liberty and licentiousness.

In all likelihood only a minority of universities, committed by their statutes and strengthened by their loyalty to those statutes or hallowed mottos, will teach not only information but also meaning, that is, true integral humanness. I hope and pray that this University will never be ashamed of a reference in its statutes to Jesus Christ as a unique revelation of God. I am saying this not only as a theologian but also as a historian of science. Only the professedly rationalist and materialist historians of science keep ignoring that belief in creation out of nothing played a pivotal role in the formulation, in the fourteenth century, of some crucial notions anticipating and preparing Newtonian physics. It is still to gain broad awareness that without the dogma of Incarnation, according to which only the Son is begotten and therefore the world cannot be a begetting or an eternal, necessary part of God, the dogma of creation out of nothing would have lost much of its incisiveness.[61] But history of science or not, the incomparable fact of Christ demands the careful study on the part of all those who in this age of science put the highest premium on unconditional respect for facts. Such a respect is easier to say than to implement. I wonder whether T. H. Huxley ever did with respect to Christ what he urged with respect to any fact of Nature: sit down before fact like a child and follow its voice no matter where it may lead.[62]

Teachers at universities where integral humanness is the supreme standard will have plenty of opportunities to take heart from the slips of tongue all too frequent in universities where hallowed mottos (*Novum Testamentum-Vetus Testamentum, In numine Dei viget,* etc. . . .) have degenerated into mere slips of tongue. Such a slip of tongue is a remark of Prof. Wilson that religion cannot be eliminated because like everything else, it too is a matter of genes in whose subservience we allegedly all are.[63] Such a slip of tongue is that popular Harvard course, which has been dubbed "Guilt 33" by the undergraduates.[64] Its two teachers organized it under the impact of having discovered that children, never exposed to religion, all too often develop a sense of guilt in a genuinely moral sense. Such slips of tongue are indicative of that sheepish manner in which a dehumanized science reinstates man, the religious animal, and bows to the truth of *anima naturaliter Christiana.* Thus, when teachers in universities committed to meaning feel downcast by their minority status, let them take heart from the fact that they will be the last consistent defenders not only of the humanities but also of true science. Or as Chesterton put it at the end of his *Heretics*:

> We shall be left defending not only the incredible virtues and sanities of human life, but something more incredible still, this huge impossible universe which stares us in the face. We shall fight for visible prodigies as if they were invisible. We shall look on the impossible grass and skies with a strange courage. We shall be of those who have seen and yet have believed.[65]

[1]John Henry Newman, *The Idea of a University Defined and Illustrated* (8th ed.; London: Longmans Green and Co., 1888), p. 145.

[2]The remark, probably made in December 1871 at a meeting in New York of Williams alumni, provided the title for R. Frederick's book, *Mark Hopkins and the Log: Williams College, 1836-1872* (New Haven: Yale University Press, 1956).

[3]Letter of April 11, 1892, to Prof. E. Ray Lankester, in *The Life and Letters of Thomas Henry Huxley,* by L. Huxley (New York: D. Appleton, 1901), vol. II, p. 328. The "factory of knowledge" stood in contrast to the "storehouse of knowledge," or the medieval university.

[4]A favorite remark of Sir George Pickering, Regius Professor of Medicine at Oxford University.

[5]For a long list of no less shocking topics of theses that earned the Ph.D., see A. Flexner, *Universities: American, English, German* (1930; reprinted

with a new introduction by C. Kerr; New York: Oxford University Press, 1968), pp. 153-54.

[6]Unidentified author, quoted in *A New Dictionary of Quotations on Historical Principles from Ancient and Modern Sources,* selected and edited by H. L. Mencken (New York: Alfred A. Knopf, 1946), p. 1238.

[7]Ibid.

[8]*The Idea of a University,* p. 145.

[9]Thomas Carlyle in his fifth lecture, "The Hero as Man of Letters. Johnson, Rousseau, Burns," delivered on May 19, 1840, in the series, *On Heroes, Hero-Worship and the Heroic in History.*

[10]Newman would find even more revolting the custom of taking these words of President Wilson out of context. In his speech given before the Y.M.C.A. of Pittsburgh on October 24, 1914, Wilson started with a reference to his experience as president of Princeton University that students far from being "arch radicals" were markedly conservative, resisting any change relating to instruction and campus life. Worse, they were most unwilling to admit their strong attachment to their fathers' outlook, limited in most cases to their business or profession. Hence the need, stressed by Wilson, to expose young men to the far and wide world of ideas and make them "as unlike their fathers as possible." The main thrust of Wilson's speech is hardly ever quoted. Indeed, presidents of Christian colleges and universities would for the most part be very unwilling to insist nowadays on an atmosphere in their campuses which Wilson believed to permeate Y.M.C.A. clubs. After their initial hostility to Y.M.C.A., most churches realized, Wilson stated, that "it was a common instrument for sending the light of Christianity out into the world in its most practical form, drawing young men who were strangers into places where they could have companionship that stimulated them and suggestions that kept them straight and occupations that amused them without vicious practices; and then, by surrounding themselves with an atmosphere of *purity and of simplicity of life,* catch something of a glimpse of the great ideal which *Christ lifted* when he was elevated *upon the Cross.*" Quotations are from the text of Wilson's speech, printed by the U.S. Government Printing Office, in a ten-page-long pamphlet, No. 67334-14. See especially pp. 3 and 4-5 (italics added).

[11]A remark in Henry Fielding's comedy, *The Temple Beau* (1730); see *The Works of Henry Fielding* (new ed.; London: J. Johnson, 1806), vol. I, p. 192. Fielding, much better remembered for his *Tom Jones,* would find many more tragic than hilarious incidents in the mores prevailing in most modern campuses.

[12]In the sense, of course, that natural theology provided a rational justification of the possibility of revealed truth. Newman, who delivered in 1852 the nine lectures forming *The Idea of a University,* stressed the intellectual

indispensability of natural theology in the face of fideism which he knew all too well from his Protestant years and which also made heavy inroads into Catholic theology during the decades of Romanticism. Newman's insistence on natural theology was a prophetic anticipation of the Thomistic revival stressed by Leo XIII and his successors. While the demise of natural theology, during the last two decades, in most Catholic colleges and universities would have dismayed him, he would see it as a logical consequence of the betrayal there of the philosophical ideas of Thomas Aquinas.

[13]S. Weinberg, *The First Three Minutes: A Modern View of the Origin of the Universe* (London: André Deutsch, 1977), p. 154.

[14]References are to the Bantam Book edition (New York, 1979). See pp. 3 and 217.

[15]Ibid., p. [ii].

[16]Ibid., p. xi.

[17]See Max F. Schulz, *The Poetic Voices of Coleridge: A Study of His Desire for Spontaneity and Passion for Order* (Detroit: Wayne State University Press, 1963). Schulz marshalled strong evidence against the view that at least some works of Coleridge were the products of "manual somnambulism" (p. 7).

[18]A. N. Whitehead, *The Function of Reason* (Princeton: Princeton University Press, 1929), p. 12.

[19]A. Einstein, "Physics and Reality" (1936) in *Out of My Later Years* (New York: Philosophical Library, 1950), p. 59.

[20]For details, see ch. 12, in my *The Road of Science and the Ways to God*, The Gifford Lectures, 1974-75 and 1975-76 (Chicago: University of Chicago Press, 1978).

[21]Reported by E. Gilson in his *D'Aristôte à Darwin et retour: Essai sur quelques constantes de la biophilosophie* (Paris: Vrin, 1973), p. 49.

[22]A very old truth, given new popularity in our time by Sir Karl Popper, although his philosophy can hardly accommodate a science of cosmology properly so-called.

[23]For details, see my *The Road of Science*, pp. 121-22.

[24]It was Kant's exposure to Rousseau's *Julie, ou la nouvelle Héloise* in 1762 that crystallized his groping for that autonomy. For details, see my book *Angels, Apes and Men* (La Salle, Ill.: Sherwood Sugden, 1982); ch. 1.

[25]A vast documentary evidence on behalf of this claim is given in the introduction and notes to my translation of Kant's cosmogony, *A Universal Natural History and Theory of the Heavens* (Edinburgh: Scottish Academic Press, 1981).

[26]See *Angels, Apes and Men* (ch. 3) and my earlier publications, *The Paradox of Olbers' Paradox* (New York: Herder and Herder, 1969), pp. 158-60 and "Das Gravitations-Paradoxon des unendlichen Universums," *Sudhoffs Archiv* 63 (1979), pp. 105-22.

[27]Not even K. Schwarzschild's two discussions, in 1901 and 1908, in terms of four-dimensional geometry, of the relation of the total mass, supposedly confined to the Milky Way, and to the empty space surrounding it. See my *Milky Way: An Elusive Road for Science* (New York: Science History Publications, 1972), p. 276.

[28]For details, see E. Gilson, "Theology and the Unity of Knowledge," in L. Leary (ed.), *The Unity of Knowledge* (Garden City, N.Y.: Doubleday, 1955), pp. 40-42.

[29]For details, see the introduction to my translation of Bruno's *The Ash Wednesday Supper: La cena de le ceneri* (The Hague: Mouton, 1975).

[30]For details, see *Universal Natural History and Theory of the Heavens*, pp. 280-81.

[31]That the law of the instability of the homogeneous, the primary principle in Spencer's cosmogony, never appeared to him intrinsically contradictory, may have its explanation in his candid admission, "I am never puzzled." See *Autobiography* (New York: D. Appleton, 1904), vol. I, p. 462.

[32]This is certainly true of the Laplacian version of that nebula and of the popular or cliché-ridden version of Kant's dicta on the subject. Actually, Kant in his cosmogonical work carefully specified the initial condition, which in his belief gave rise to the actual shape of the universe.

[33]As can readily be gathered from his notebooks composed in 1837-38. For details see *Angels, Apes and Men,* ch. 2.

[34]Part of that story is the Marxist espousal of the doctrine of eternal recurrence. See my *From Eternal Cycles to an Oscillating Universe* (Edinburgh: Scottish Academic Press, 1974), pp. 312-18.

[35]See my *The Paradox of Olbers' Paradox,* ch. 9, and *The Milky Way,* ch. 8.

[36]On some apt comments on the significance of zero, see my article, "The Chaos of Scientific Cosmology," in *The Nature of the Physical Universe: 1976 Nobel Conference,* organized by Gustavus Adolphus College, edited by D. Huff and O. Prewett (New York: John Wiley, 1979), p. 92.

[37]Especially, pp. 102-08. For somewhat different figures, see B. Lovell, *In the Centre of Immensities* (London: Hutchinson, 1979), pp. 120-26.

[38]Such an inference can be made on the basis of the finding which earned to V. Fitch and J. Cronin the Nobel Prize, that for two out of every thousand artificially induced decays of neutral K_2 mesons, symmetry laws do not hold.

[39]For an early statement, see B. Carter, "Large Number Coincidences and the Anthropic Principle in Cosmology," in M. S. Longair (ed.), *Confrontation of Cosmological Theories with Observational Data* (Dordrecht: D. Reidel, 1974), pp. 291-98.

[40]M. B. Wise, H. Georgi, and S. L. Glashow, "SU(5) and the Indivisible Axion," *Physical Review Letters* 47 (August 10, 1981), p. 402.

[41]The philosophical shallowness of that book was held up to a well-deserved

criticism by L. S. Stebbing, *Philosophy and the Physicists* (London: Methuen, 1958).

[42]For the first time in my *The Relevance of Physics* (Chicago: University of Chicago Press, 1966), pp. 127-30.

[43]See Letters of Jan. 1, 1951 and March 30, 1952 in *Lettres à Maurice Solovine* (Paris: Gauthiers-Villars, 1956), pp. 102 and 115.

[44]Students are hardly ever made aware of the true thrust of Carl Sagan's studied reliance on the use of triple and at times quadruple negatives.

[45]Quoted in M. Link, *Take off Your Shoes* (New York: Paulist Press, 1974), p. 40.

[46]Words of B. Murchland, quoted in *Newsweek* (October 13, 1980), p. 113. The report itself, *The Humanities in American Life* (Berkeley: University of California Press, 1980), is noteworthy for its sedulous avoidance of any philosophical question that may arise in connection with man and mankind.

[47]I have in mind studies prepared recently at the University of Aberdeen by Professor P. Wilkinson. See especially his "Proposals for Governments and International Responses to Terrorism," in P. Wilkinson (ed.), *British Perspectives on Terrorism* (London: George Allen & Unwin, 1981).

[48]The phenomenon is not at all new. What is new is the inability of an ever larger number of students to articulate, however simply, their intensely felt need for meaning. Quite a difference with respect to the 1950s when 78 percent of almost 8,000 students selected from 48 colleges and universities stated in a survey conducted by social scientists that their main goal in education was the finding of meaning and purpose to life. See V. E. Frankl, *Man's Search for Meaning: An Introduction to Logotherapy* (New York: Simon and Schuster, 1963), p. 55.

[49]*Le Monde* (Aug. 13, 1982), p. 5, from an interview with R. Leroy, grandmaster.

[50]J. S. Mill, *Autobiography,* with an introduction by C. V. Shields (Indianapolis: Bobbs-Merrill, 1957), p. 153.

[51]G. K. Chesterton, *Heretics* (London: John Lane, 1905), p. 13.

[52]Ibid.

[53]Mill, *Autobiography,* p. 145. That Mill speaks actually of mathematical and physical sciences, should pose no problem. Nor should his lumping of theists with idealists; Mill merely confused metaphysics with idealism.

[54]*Heretics,* p. 13.

[55]Ibid., p. 51.

[56]*On Human Nature,* p. 200.

[57]As I argued in my "Chance or Reality: Interaction in Nature and Measurement in Physics," reprinted here as Ch. 1.

[58]As acknowledged by Prof. Wilson, *On Human Nature,* pp. 1 and 200.

[59]The number of such astronomers and cosmologists is much larger than suspected.

[60]A claim memorably stressed in his debate with Fr. Copleston on the existence of God.

[61]A discussion and documentation of this point can be found in my *Cosmos and Creator* (Edinburgh: Scottish Academic Press, 1981), pp. 65-79.

[62]"Sit down before fact as a little child, be prepared to give up every preconceived notion, follow humbly wherever and to whatever abysses nature leads, or you shall learn nothing." *Life and Letters of Thomas Henry Huxley*, vol. I, p. 219. These words are part of a letter (Sept. 23, 1860), in which Huxley declined the suggestion that following the death of his seven-year-old son, he should consider the message of Christ about immortality. One wonders whether the thought had crossed Huxley's mind that what he urged with respect to facts, Christ most emphatically urged with respect to Himself when He singled out children as the ones possessed of the proper attitude toward His Kingdom. At any rate, it was typical of Huxley, the natural scientist, that for him the realm of facts did not really include the very real facts of human history.

[63]*On Human Nature*, p. 42.

[64]The official entry for the course is "Literature and Arts A-20" and has the title, "The Literature of Christian Reflection," taught by Dr. Robert Coles and Prof. Robert Kiely. The course seems to offer much more to young men and women looking for meaning, than, for instance, the course described in the catalogue of Harvard Divinity School as "Applied Theology 142. Sociology of Religion and Teaching. This is a macro/micro humanistic approach to the sociology of religion and teaching containing four components: (1) Examination of historical paradigms; (2) Field studies of radical sociological change in religious states; (3) Analysis of social-cultural continuities and discontinuities, derived and underived religious posturing, dimensions of tension, crisis confluence and synthesis in cross-cultural states; (4) A correlation of sociology and teaching enabling the student to distill transcendent processes as well as to develop a more cogent language of religion in the context of educational goals." To this the best reaction still is *The New Yorker*'s comment: "Those who pass, go straight to Heaven" (Feb. 26, 1979, p. 103, col. 1).

[65]*Heretics*, p. 305.

11

The Greeks of Old
and the Novelty of Science

In speaking of science, it is hardly possible not to think of the Greeks of old. The phrase, coined over half a century ago, that "science is the Greek way of thinking about the world,"[1] merely codified an already hallowed conviction, which today is more robust than ever. The reason for this lies in geometry, the highest achievement of ancient Greeks in science and the most enduring form of their commitment to the ideals of reason. Today, scientific rationality is bound up with a mathematics cast in increasingly more esoteric forms of geometry. Ever since Planck and Einstein ushered in a new age of physics, major advances in our understanding of the empirical realm were so many witnesses to what was aptly called the "unreasonable effectiveness" of pure mathematics in physical science.[2] The universe, which for Newton appeared to

Reprinted with permission from the Festschrift in memory of Konstanin I. Vourveris (1902-1981), Professor of Greek at the University of Athens and Founder-President of Hellenic Society for Humanistic Studies (Athens, 1983), pp. 263-77.

be a huge machine, is seen by leading twentieth-century physicists as a pattern in numbers.[3] In reflecting on this, modern physicists find it natural to refer to Plato and to his bold vision of a universe built of the five perfect solids.[4] Distant as Plato's geometrical account of the five elements may be from the equations of modern physics concerning elementary particles, it anticipated the resolve of physicists of our times to make their equations more effective, that is, mathematically more beautiful.[5]

The indebtedness of modern science to the Greeks of old is, of course, far more detailed and specific than can be indicated by a general perspective, the viewing of the universe as a geometrical pattern. The frustration of modern physicists to find the truly indivisible parts of matter is a replay of the inconclusive dispute very much alive in Plato's time whether matter was continually divisible or made up of atoms, that is, parts that could not be cut any further. As to the realm of the very large, the Greeks had already discoursed on questions, much controverted today, whether the universe is finite or infinite, one or many, truly ordered or basically haphazard. They could also give answers to particular questions with an exactness which is satisfactory even by modern scientific standards. Their estimates of the radius and circumference of the earth, of the size of the moon, and of the earth-moon distance are so many evidences of impeccable scientific reasoning and of experimental accuracy. As to their discovery of the precession of the equinoxes it remains one of the greatest discoveries of all times. To the analysis, which Hero of Alexandria gave of the five simple machines (wheel and axle, lever, system of pulleys, wedge, endless screw), little if anything was added during the next two millennia. But even when the Greeks of old could not rely on geometry, whose effectiveness did not at all seem to them "unreasonable" or surprising, some of them could come up with scientific attainments which astonish by their modernity. The booklet that Plutarch wrote on the moon[6] has a breathtakingly modern ring in its matter-of-fact passages on the earthlike properties of a celestial body which only the telescope could finally divest of reputedly supramundane qualities. Linnaeus and Cuvier, two giants of modern science and long the heroes of Darwin, became mere schoolboys in his estimate after he had discovered Aristotle, the biologist.[7] At any rate, Euclid has been, until recently, a household word even with schoolboys.

Such a recital could go on for long, and the longer it continued the more revealing it would appear. Even in nutshell accounts of the history of science the Greeks of old represent the first significant chapter. Authors of longer and sufficiently documented science histories invariably convey their frustration owing to the limits of space they can devote to the Greek miracle in science. The variety and richness of topics in anthologies of Greek science may give even to a layman the impression that science is indeed the Greek way of thinking about the world. Indeed, their science, from the Ionians on, always had relevance for the world as a whole. Here too they had anticipated modern times in recognizing the elementary truth that ultimately all science is cosmology.[8] Ptolemy, the last towering figure of ancient Greek science, is best remembered as a cosmologist, although he excelled as a geographer and an algebraist as well. The value of his *Almagest* may be conveyed by a mere recall of the fact that Copernicus' *De revolutionibus* did not represent an immediate improvement over Ptolemy's work concerning the prediction of the position of the sun, the moon, and the planets.

That the heliocentric system of Copernicus contained major and crucial advantages over the geocentrism of Ptolemy for the study of planetary motions became clear but gradually. This story, leading through Kepler, Galileo, and Horrocks to Newton, is too well known to be outlined here. It is the story of the rise of a science in which one discovery generates another and makes science a self-sustaining, open-ended venture. What may appear to be a mirror image of this story is much less known and is hardly ever discussed. Yet a discussion is very appropriate, and partly because of the four hundred years which separate our decades of space probes and nuclear energy from Copernicus who, in a sense, started it all. Those four centuries seem to modern eyes as replendent in ever greater scientific glory. The very opposite impression is given by the four centuries preceding Ptolemy, although at their beginning, or about 250 B.C., one finds none other than "The Ancient Copernicus," the title given by Sir Thomas Heath, a leading authority on Greek mathematics, to his still classic monograph on Aristarchus of Samos.[9]

In all likelihood, Aristarchus not only spoke of the sun as the center of planetary orbits but also worked out to some extent, though not nearly as much as did Copernicus, the heliocentric motion of planets. His treatise on the sizes and distances of the earth, moon, and the sun[10] shows him both an original mind and accomplished geometer, who could have easily coped with the problem of giving a geometrical form to the intricacies of planetary motions as taking place around the sun. But would that form, however elaborated, have given heliocentrism a chance in the Hellenistic world, and triggered, 1800 years ahead of time, a most momentous development with Aristarchus as an anticipation of Copernicus and with Ptolemy as an anticipation, if not of Einstein at least of Newton? This question, so speculative if not idle in appearance, may be given a not too speculative answer. Although the extant record is meager concerning reaction to Aristarchus' heliocentrism, there is no indication that it prompted debate in a world eager to debate anything. Only one scholar, Seleucus, is known to have endorsed it about a hundred years after Aristarchus.[11] In general, reaction was negative and at times hostile. The source of that hostility was the belief sanctioned by Plato and Aristotle, and widely entertained even outside the circles of their followers, that the corruptible earth represented a realm very different from the one between the moon and the fixed stars. Indeed the two realms were believed to be so different as to be governed by two sets of very different laws corresponding to two sets of very different kinds of matter. Their difference was nothing less than that between an ordinary and a divine existence. To that belief Plutarch's idea, according to which the "divine" moon was an earth-like body, was just as sacrilegious as was Aristarchus' idea that the "ordinary" or corruptible earth moved in the "divine" or incorruptible realm of the planets and stars. Even the idea entertained by the Stoics, that the earth's position was determined by its equilibrium with respect to the air surrounding it,[12] and not by its function as the hearth or center of the universe, was hardly reconcilable with that belief. All these three ideas are mentioned in the same breath by Plutarch who laughs away the problems they raise. He does so in his book on the moon and in a context which includes the often quoted passage: "Cleanthes thought that the Greeks ought to lay an action for impiety against Aristarchus the Samian on the ground that he was disturbing the hearth of the universe because he sought to save the phenomena by assuming that the heaven is at rest while the earth is revolving along the ecliptic and at the same time is rotating about its own axis."[13]

The passage may help in understanding Ptolemy's rejection of heliocentrism as "ridiculous" and "absurd."[14] The passage reveals something about Plutarch, bold enough to turn the moon into an earthlike body, but not bold enough to let the earth move around the sun. The passage is also noteworthy for its phrase, "to save the phenomena," which, as will be seen later, became the capsule formulation of the majority view among the Greeks as to what science aims at. Most importantly, the passage is as good a clue as any ever to be found to the question, apparently so speculative, as to why heliocentrism remained a barren proposition within Greek science. This question, by conjuring up the vista of something intrinsically antiscientific or irrational in Greek thought—often and indiscriminately taken for the embodiment of sheer rationality—questions the very view that science is the Greek way of looking at the world. For, if the Greek way of looking at the world had been so genuine an expression of rationality, then science, which is certainly a rational investigation, would have hardly failed to reach already in the Greece of old the stage of a self-sustaining enterprise.

Ancient Greek science came much closer to that stage than was the case with science in all other ancient cultures—China, India, Egypt, and Babylon.[15] Although all these cultures could boast of some evidences of scientific ingenuity (e.g., the magnet, printing, and gunpowder in China; the decimal system of counting in India), in none of them had a systematic development of any area of science taken place. Compared with ancient Chinese, Hindu, and Babylonian astronomies, if they can be called such at all, ancient Greek astronomy appears to be a full-fledged science. No different is the case when algebra and geometry are considered. In all these fields ancient Greek science not only transcended the stage of careful description, observation, and classification, the stage exemplified by Aristotle's biology (or Galen's medicine), but moved up to the stage where the entire body of knowledge was a derivative of some fundamental postulates. Yet, whatever the excellence of Euclidean geometry and Diophantine algebra, they strangely seem to be thwarted in their development. This is even more the impression one gains about ancient Greek astronomy. The failure of the Greeks of old to invent the telescope is largely irrelevant. The moment of truth for astronomy did not come with the telescope of Galileo, but with Newton's insight, not at all dependent on the use of the telescope, that the fall of an apple on the earth and the constant bending of the moon's orbit toward the earth were two cases of one and the same process. It was an insight that did not even depend on the application of infinitesimal calculus toward which Archimedes made halting steps though falling far short of what came out from the hands of Newton and Leibniz.

Newton could not have developed that crucial insight into what had become the science of the *Principia* and all modern science built on it without taking full advantage of infinitesimal calculus. This must not, however, distract from what was truly crucial in his insight. It concerned the fusion of celestial reality and earthly reality into one with the help of mathematics (geometry). Such was the feat which the Greeks of old failed to achieve. Since their observations of celestial motions, and of the moon in particular, were remarkably accurate and since their geometry was marvel itself, the source of their failure must be sought in their view of physical reality, or of the world in short. In other words, in a sense touching far more concretely on physical reality than would be suspected, there must have been something very defective in the way in which the Greeks of old looked at the world.

Such a suspicion cannot be sensed in the few cases when historians and philosophers of science consider a very baffling aspect of the history of science in ancient Greece: a brief outburst of creativity (450-350 BC) was followed by a long period of ingenious elaborations (350 BC-150 AD) leading to a protracted stagnation which came to an end around 600 AD. Marxists are wont to ascribe the cause of this outcome to social conditions, and above all to the reliance of Greek society and economy on slavery.[16] Their contentions are just as artificial and impossible to apply to other phases of science (witness the stifling of scientific creativity in modern Marxist regimes), as is another and somewhat older contention which blames Christianity for the stifling of science and rational thinking in classical antiquity. Whatever the merits of the claim that Christianity fostered irrational mysticism and contempt for cultural improvement, Greece had long before become the eager recipient of Eastern—mostly Babylonian and Egyptian—astrology, soothsaying, and haruspicy. Hellenistic science and philosophy had for long been stagnant when, about the time of Constantine, Christians became a minority to reckon with. The explanation made popular in our time by Alexandre Koyré that while it is possible to explain why science could not develop in the heavily bureaucratic and rigidly autocratic societies of China and Babylon, it is not possible to explain why science developed in Greece,[17] is tantamount to writing off a glaring problem before thoroughly examining it.

A hundred years before Koyré, William Whewell suggested that the Greeks failed to make enough progress in science because they were unable to match their observations with appropriate ideas.[18] As an author of several large works, which set the standard for over two generations on the history and philosophy of science, Whewell could have been expected to dwell on his suggestion which after all points in the right direction. The matching of observations with appropriate ideas is a question of knowing reality. A close and sustained look at the Greeks' failure to achieve that matching in a way productive of science should therefore seem of prime importance. Certainly such should be the case not only with those, who with Burnet look at the development of the philosophy of the Greeks of old as a process dependent on their progress in science, but also with those, whose number is even larger, who are aware of the crucial role which science came to play in human history.

They all know that the Greeks of old made many observations and had as many if not even more ideas. But while particular ideas are many, the general ideas of "looking at the world" are few. In every great culture one of those few general ideas becomes preponderant and functions as the chief determinant of the manner in which the world is looked at. The matching of particular observations with particular ideas is done within that general world view. The success of that process, which is also the success of science in a particular culture, cannot therefore be evaluated without a close investigation of that world view and also of its presence in scientific theories bearing on this or that particular aspect of physical reality.

It tells something of the academic world in general and of academics in particular working in the history and philosophy of science, that although a very crucial part of that investigation had early in this century been presented and supported with massive documentation, it largely remained a closed book. This should seem all the more strange because books on Greek science mostly fail to match even in extent, to say nothing of richness of content, the first two volumes of Pierre Duhem's *Le Système du monde*.[19] Its other eight volumes, no less massive, remained an equally unexplored storehouse of information on the science of almost ten medieval centuries.[20] As to the first two volumes, dealing with the history of cosmological ideas from Plato to Proclus, they would have done credit even to a classical scholar, let alone to that theoretical physicist which Duhem considered himself above all.

In view of Duhem's enormous output in theoretical physics, the extent and originality of his work in the philosophy and history of science may seem to call for another lifetime and all the more so as he died at the relatively young age of fifty-five in 1916. In addition to limitations of time there is perhaps the even more important limitation in versatility. Mastery of any of the exact sciences is hardly ever accompanied with a sense of history. Duhem was a born historian—he almost chose to study history instead of physics[21]—and a linguist, to say nothing of his phenomenal memory. Two decades had gone by following his graduation in 1882, at the age of 20, from the Collège Stanislas in Paris, when he published his first study of Greek science, but he did not have to relearn Greek. As a former classmate of his noted, Duhem only had to "remember" what he learned as a schoolboy.[22] His memory stood him in good stead because no sooner had his attention turned around 1903 to a vigorous study of the history of mechanics as a means of probing its theoretical foundations, than Greek science began to be a steady feature in his publications. His *Evolution de la mécanique* started with a chapter on the Greeks[23] and so did his *Les Origines de la statique.*[24] The remark in the latter that the medieval renaissance of the study of statics owes more to the ideas of Aristotle than to Archimedes[25] was an evidence of his sensitivity to philosophical nuances and their historical provenance. His penetration into the philosophical background of Greek theories on the notion of physical science was in full evidence in his classic volume on the philosophy of physics, published in 1906[26] and even more so in his monograph on the evolution of the notion of physical theory from Plato to Galileo, published in 1908 under the title, ΣΩΖΕΙΝ ΤΑ ΦΑΙΝΟΜΕΝΑ.[27]

None of these publications did, however, suggest the eventual appearance of that rich and penetrating presentation of the fundamental principles governing Greek scientific thought about the cosmos which is contained in the first two volumes of *Le Système du monde.* This is not to suggest that those volumes are void of chapters which would do credit to any author who restricts his view according to the positivist rules of historiography to what proved to be technically useful for later phases of science. Duhem's chapters on the precession of the equinoxes, on the dimensions of the world, on epicycles and deferents still are a major source of information. But "positivist" as Duhem could be, he was a very peculiar one, a point which is amply evidenced in the major strokes with which he painted the vast panorama of Greek science. Had he been a positivist in the customary sense, Duhem would have hardly filled over 800 more large octavo pages after declaring that "since the century of Pericles, Hellenic thought conceived with admirable clarity that form of science which today we call theoretical or mathematical physics; it understood how the geometer can pose at the beginning of his investigation a small number of simple and precise hypotheses, how with the help of deductions he can erect on those foundations a system able to save all the phenomena which sensory perception recognized by observing the data of nature."[28] After that any further extensive discussion could only seem anti-climactic even if it bore out the contentions and anticipations of positivism.

Duhem was a positivist with qualifications that made him a metaphysical realist. That he was not fully aware of this fact makes his writings on the philosophy and history of science a fascinating study. What he wrote on Greek science is particularly instructive in this respect. The positivist rules of historiography seemed to be obeyed by Duhem's remark with which he justified the detailed attention given by him to theories, such as absolute space and motion, in all appearance so many excessive metaphysical reifications of abstract extrapolations: "We do not discuss here this question, because we do not want to do the work of a philosopher but of a historian. Now, it is enough for a historian, who accords importance in his exposition to the doctrine of absolute space and motion that at very different epochs very great minds proposed it."[29]

As is well known, positivist historians of philosophy and science do not always obey this positivist precept. They readily neglect all that part of intellectual history which is not a direct anticipation of their antimetaphysical and antirealist countermetaphysics. Duhem had, however, already stated in a memorable context in 1906 that for him the history of physics was an evidence that its succeeding stages were so many advances toward an ultimate comprehensive theory which best corresponded to physical reality.[30] In an even less positivist vein he saw the hand of divine Providence behind such development.[31] Unfortunately, he never felt the need to articulate his implicit espousal of metaphysical realism with that finesse and historical richness which he displayed time and again about any other topic. He remained a disciple of Pascal, stressing the act of intuition, if not faith, as the sole though indispensable avenue from mathematical theory to reality.[32] While such a position could be articulated impressively (Duhem was a master of style), it deprived him from seeing to the philosophical rock-bottom of the failure of Greek science. Had he seen it, he would have spoken of it in great detail because he could have done so as a historian. Of all the dialogues of Plato he failed to appreciate the one which, as will be seen later, far more than all the others made history. His failure to do so is all the more remarkable because Duhem the historian did much better justice to the philosophical consideration everywhere dominating Greek science than did its other historians, before and after him. His view that the Greeks had formulated as early as the times of Pericles the notion of modern theoretical physics, was not reached by slighting, as positivists would have done, the theological framework of knowledge within which Plato inspired the scientific methodology of "saving the phenomena."[33] Again, Duhem payed vast attention to Aristotelian physics and especially to its philosophical foundation, which is the vindication of the primacy of sensory perception against the abstract ideas of Plato. Clearly, no positivist would have done so if he had foreseen, as Duhem did, that he would have to declare after more than a hundred pages, full of Greek quotations, that the system of Aristotle is a "monument which has the unshakeable solidity of a block (of granite) and the purity of lines of the most beautiful work of art." Curiously, while Duhem mentioned physics as part of that monument, he added in the same breath: "Of the physics of Aristotle no stone on stone is left. Modern physics, in order to put itself in the place of Aristotelian physics, will have to demolish successively all its parts; undoubtedly, many a fragment borrowed from that antique monument will be retained in order to build the walls of the new edifice; but before finding its place in this new construct for which it was not cut, it will have to receive a form altogether different from what once it displayed."[34]

Duhem then specified the necessarily circular motion of the heavenly regions and the necessarily "natural" motion to "natural" places in the sublunary regions as the two tenets which are the basis of Aristotelian physics and the causes of its collapse. Of these two causes Duhem viewed with obviously greater fascination the former, the eternity of necessarily circular motions of the heavenly bodies. He had good reasons, both philosophical and historical, for doing so. Philosophically, the heavenly motions are also cases of natural motion, though not *toward,* but *in* a natural place. In the heavens the goal, the target of purpose, is always achieved. Historically, even though the Stoics and the Epicureans rejected the Platonic-Aristotelian doctrine of the incorruptibility of the heavens, they retained the Platonic-Aristotelian belief that all processes on earth are a function of the heavenly motions. In keeping this perspective in focus, which is also that of the Great Year, for the rest of his account of Greek science, Duhem did justice to the importance of the general world view. He was also right in emphasizing the fusion of that view right at the beginning of Hellenistic times with astrology. Unlike other historians of Greek science, Duhem made a pointed reference to the fact that Theophrastus, Aristotle's successor at the Lycaeum, was the author of a book on the signs of the Zodiac.[35] The book was symptomatic of the torrential influx, around the time of Alexander the Great, into Greece of astrology and related arts of which beforehand there was hardly any sign. While the Platonic-Aristotelian doctrine of eternal recurrence (either specific or generic) must have cast a pallor, though still fairly rational, the same doctrine when coupled with astrology could only be the source of rank subjectivism and irrationality.

How this development evidenced itself as post-Aristotelian Greek science was running its long course was illustrated by Duhem with painstaking detail in a chapter by far the longest in those first two volumes of the *Système du monde.*[36] That chapter shows him at his best in rounding up evidence from the most unexpected corners[37] and in building his discourse up to a grand conclusion. The chapter in question is all the more significant because its topic, the tides, gave to the Greeks a palpable glimpse into the interconnection between the motion of a celestial body (the moon) and that of a terrestrial mass, the sea waters. The interconnection, hardly obvious in most parts of the Mediterranean, had not come within the ken of the Greeks until Alexander the Great's armies reached the shores of the Indian Ocean and Pytheas explored, somewhat earlier, the northeast shores of the Atlantic. A century later, as Strabo relates, the connection between the oceanic tides and the moon's motion was explicitly made by Eratosthenes, who also recognized the moon to be the cause of the much smaller variations of sea level in the Mediterranean. Strabo does not hint as to what kind of connection was meant by Eratosthenes. As the one who is best remembered for his determination of the size of the earth, Eratosthenes may very well have had in mind a connection that could lend itself to a scientific elaboration in the modern sense. That the Greek mind was capable of pursuing such a line of thought concerning the tides (and made thereby a big step toward the ideal of Newtonian or mechanistic physics) is strongly suggested by the subsequent speculations of Seleucus, who flourished about a hundred years after Eratosthenes. A supporter of Aristarchus' heliocentrism, Seleucus saw in oceanic tides the perturbation exerted by the moon's motion on the atmosphere of a rotating earth![38]

Just as Seleucus was a lonely figure in his support of heliocentrism, his purely mechanistic idea of the tides also failed to find a favorable reception. Rather, the obvious connection of the moon's motion with the tides was taken by most Hellenistic men of science and philosophy as a major evidence on behalf of astrology. The data, for the most part very accurate, on the various periods of tides are not the only details given by Posedonius on the subject. He also describes the moon, in the style of astrologers, as the cause, on account of her feeble heat, of all swelling, fertility, and humidification on earth, and ascribes a special role in this respect to those signs of the Zodiac which have a similar nature. In doing so Posedonius not only expressed a view to which his age was most receptive. His prominence as a Stoic author also made that view authoritative for the rest of antiquity and beyond. The result was the nipping in the bud of the significance of an idea which could have very effectively steered the development of Hellenistic science in the right direction.

The tides as a major evidence of an interaction between the heavenly and earthly regions in terms of sympathy, affection, and other patently non-quantifiable concepts, are ubiquitous in Hellenistic writings, scientific as well as other. A chief example is the *Tetrabiblos* (still the Bible of astrological lore) by Ptolemy, who in another book of his on the cosmos assigned the harmonious motion of planets to their ability to co-ordinate their movements in the manner of a group of dancers or of soldiers.[39] Another major example is Marcus Manilius' *Astronomicon,* a work which is all the more significant because its author unfolds in full the reasoning which underlies the foregoing account of tides. According to Manilius, the world is a huge personal entity, God himself, whose cosmic members are in struggle with one another, a process which is the cause of the inevitability of conflicts on earth, be they conflicts of physical forces or of human beings. Being locked up in a cosmic personalism as unpredictable as are not a few personal activities, man could hardly feel encouraged to undertake their rational investigation, let alone to aim at their control. Man at best could read the celestial signs, so that the decrees of Fate would not find him completely unprepared.

The reading of those signs obviously is as far from being scientific as is the interpretation of the feathers and voices of birds, or of the direction in which they fly. Yet throughout Hellenistic times not only the obscurantist Gnostics saw in such readings a basic source of information but also thinkers of such stature as Plotinus and Proclus. The latter recalled with great admiration Theophrastus' work on the signs of the Zodiac and his great admiration for Babylonian (Chaldean) astrology, in addition to claiming that since Chaldean astrologers made no use of the precession of equinoxes in their divinations, the precession of equinoxes must not be viewed as real, the assertions of Hipparchus and Ptolemy notwithstanding![40] The bearing of such a detail goes far beyond particulars. It should bring into focus a state of mind steeped in a general world view, within which the cultivation of science was thwarted at every turn and undermined in its essential mental resources.

Duhem did not spell out such reflections, but they can hardly fail to come to mind on reading his relentless presentation of a vast evidence of which here only a few salient aspects can be recalled. Typical accounts of science in classical and late antiquity hint at most of the disastrous impact of astrology in chemical investigations or in Galen's medicine. But alchemy, as Duhem shows, was a far more obscurantist enterprise than is generally suspected. It could only be cultivated through mystical communing with the heavenly regions, because the transformation of ordinary metals into silver and gold was seen as a grafting of divine nature on things terrestrial.[41] As to medicine, one of Galen's works is on "critical days," which are astrologically propitious for physiological processes and healing procedures. That the moon plays a prominent role there, should be hardly a surprise in view of some emotive if not mystical properties which are in the end ascribed to the moon even by Plutarch, who went further than anyone in antiquity in turning the moon into a body similar to the earth. Moreover, Plutarch did so in a way which systematically discredited all essentials of Aristotelian cosmology, from the doctrine of natural places to the finiteness of the world. The resulting world picture, an infinite number of worlds in an infinite space, each a system in itself with appropriate inhabitants, astonishes with its Newtonian modernity. In fact Plutarch goes almost as far as Newton when he says that the moon's fall towards the earth is offset by the speed of its orbiting, in the same way in which a stone in a sling is kept rotating.[42] "An idea worthy of a genius!" Duhem exclaims. "The entire celestial mechanics of Newton had to emerge from it eventually. But a very precocious idea it is, much ahead of its time. . . . For long centuries it will remain as formulated by Plutarch: a grain in the state of latent life which will germinate at the moment when there will be united all circumstances required for its development and then it will issue in an admirable blossoming."[43] Clearly, less than thirty pages from the end of his monumental account of Greek science, there could seem ample justification for the remark which Duhem made not far from its start: "As we continue traversing the history of Greek astronomy, these sentiments of astonishment and admiration will not cease to increase."[44]

But fearful presentiment is no less on the increase as the same history is being unfolded. For as Duhem notes, Plutarch not only presented views on the moon with great scientific potential. He also voiced that view of the moon, so dear to astrologers, which could but undercut scientific thought. The view was part of a cosmic view in which everything on earth was ruled by an animated heavens whose basic law was the interplay of sympathy and antipathy and which was subject to the recurrence epitomized in the Great Year. As to Plutarch, Duhem may have quoted from his book on the moon a passage towards its end according to which "it must be out of love for the sun that the moon herself goes her rounds and gets into conjunction with him in her yearning [to receive] from him what is most fructifying."[45] But Duhem had already quoted many similar passages from others and could therefore turn to the surrendering of Arab and Jewish scholars to the idea of the Great Year and the system of world it embodied to arrive at his grand conclusion: "To the construction of that system all disciples of Hellenic [Platonic] philosophy—Peripatetics, Stoics, Neo-Platonists—contributed; to that system Abu Masar offered the homage of the Arabs; the most illustrious rabbis, from Philo of Alexandria to Maimonides, have accepted it. To condemn it and to throw it overboard as a monstrous superstition, Christianity had to come."[46]

Such was as clear a declaration as there could be that the failure of the Greeks of old to achieve something of which Christianity proved itself capable, was also their failure to produce science in a modern sense. Such a science was a great novelty at the time of Newton and would have been even more so had it arisen in the Greece of old because other ancient cultures fell far shorter of catching a glimpse of genuine science than the Greeks did. But clear as Duhem's declaration was and evocative as it was of philosophical and theological depths, it did not go to the very depth of the matter. Duhem never asked why the Greeks of old fell prey to the notion of the Great Year to the extent to which they did and with a Plato so much in the lead as to let the Great Year become synonymous with the Platonic Year. Though attentive to arcane details, Duhem overlooked a glaring item in Plato's cosmology, which he took for good reasons for his starting point. In that cosmology the geometrical details, which imply an anticipation of the modern notion of physical theory, are minor compared with a view according to which the universe is not a geometrical construct but a living body. Had Duhem not neglected the larger part of *Timaeus,* he would have prefixed to his long chapter on Plato not only a chapter on the Pythagoreans but also on Socrates. For not only was Socrates an immediate teacher of Plato which the Pythagoreans were not, Socrates was also the greatest and most influential teacher of all Hellenic and Hellenistic times. His influence on history was nowhere more momentous than in the history of physics, an outcome that would not have surprised Socrates at all. The chief evidence of this is provided by Plato himself. His *Phaedo* is, from start to finish, the gripping report of an eyewitness that Socrates could justify to his friends his apparently senseless self-destruction only by recourse to a new idea of physics, a physics at complete variance with the mechanistic physics of the Ionians and especially of Anaxagoras. Since in that physics all considerations about purpose and values were destroyed by the hardness of ironclad mechanism, Socrates felt that it was to be rejected outright. He proposed to replace it with a physics which explained for what purpose things moved.[47] In that physics, conjured up in the second half of *Timaeus,* where the universe is described in terms of a human body (a notion subsequently worked out with seemingly unassailable logic by Aristotle), everything was moved by purpose. Once this was taken for granted no questions could be raised about the purposefulness of Socrates' decision to drink the hemlock.

Duhem made only two insignificant references to Socrates,[48] and none to *Phaedo.* Had he paid careful attention to the message of *Phaedo* and its impact, he would have perceived that the history of Greek science is not so much a struggle to "save the phenomena" but a struggle "to save purpose."[49] A noble struggle to be sure, which modern scientific times should all the more emulate as the erosion of the sense of purpose and values is more excessive than ever, precisely because of the universal evaluation of almost everything in terms of quantities alone. For a remedy to that reductionist disease it is tempting to look to the Greeks, the masters of the Western mind in more than one way. In this case a careful recall of the experience of the Greeks of old may have remedial effects only if we learn from their mistake and especially of that of Socrates, possibly the greatest mistake he ever committed. Although unbearably relentless in questioning everything in order to reach the true and the good, he failed to ask whether truth and goodness were really served by his dramatic proposition that the cure of disease fostered by mechanistic physics as a basic philosophy called for a wholesale replacement of mechanism with teleology as a basic explanatory device. The proposal implied that there was nothing true or good in considering mechanism and quantities, a most questionable proposition, and unquestionably an extremist one. As such it was burdened by an inherent fallacy, which exacted its revenge in the exclusive attention to purpose, or the extremist stance of seeing purpose everywhere.

To maintain a balanced view through rendering their respective dues to mechanism as well as teleology should seem to be the dictate of basic rationality and, as such, hardly a difficult task to achieve. The case of the Greeks of old shows the contrary. The greatest among them opted either for one or the other extreme and, from Socrates on, mostly for the extreme of pan-teleologism. The latter received its most encompassing formulation in the cosmology riveted on the Great Year, a cosmology which was a mixture of rank subjectivism and inescapable determinism. The cause of science could not be promoted by either. For if determinism is as universal as to have man completely under its sway, then man's free investigation of the universe becomes inconceivable. That man's freedom and the contingency of things are two questions on which Aristotle's discourse becomes hurried, hesitating, and least in consonance with that common-sense experience which he had set as the touchstone of truth, should not seem surprising. Duhem's portrayal of this is far more extensive and penetrating than what is offered by most historians of Greek philosophy.[50] That historians of Greek science pass over it, as a rule, in silence should seem equally strange, since science is a free investigation on the part of man and can only be such if his object of investigation is a contingent world. Conviction on both points was secured only by Christianity for which the freedom of man is an indispensable tenet. And so is, of course, the contingency of a world which is created, a point that could but remain hidden in a world wholly subject to the subjective because ultimately it was believed to be uncreated or divine. For Plato and the Greeks, the world was not created but generated, or rather begotten from a divine substance. For Christians the only divinely begotten entity was the Son, alone consubstantial with the Father, the Creator. The world in Christian perspective had to be created, that is, contingent in the deepest sense. But since creation was the act of a rational Creator, infinitely superior to a mere demiurgos, the work of creation had to be fully consistent, that is, rational. Such was the fuller perspective of the "Word became flesh," a perspective which the Greeks of old could not muster. This is why science implies much more than the Greek way of looking at the world, a way which, however rational as long as it dealt with the abstractions of geometry, was not rational enough when it came to physical reality. In the end it became the prisoner of an irrationality which barred access to the novelty of a self-sustaining science, the only science worthy of its name.

[1]An often quoted dictum of J. Burnet. See the preface to the third edition of his *Early Greek Philosophy* (London: Adam and Charles Black, 1920).

[2]E. P. Wigner, "The Unreasonable Effectiveness of Mathematics in the Natural Sciences," *Communications on Pure and Applied Mathematics* 13 (1960), pp. 1-14.

[3]For details, see chs. 2 and 3 in my *The Relevance of Physics* (Chicago: University of Chicago Press, 1966).

[4]See, for instance, W. Heisenberg, *Physics and Beyond: Encounters and Conversations,* translated from the German by A. J. Pomerans (New York: Harper and Row, 1971), pp. 7-9.

[5]Dirac emphatically put in this light the success of his own work. Quite recently the astrophysicist S. Chandrasekhar explored this theme. For a survey and documentation, see section, "Cosmic Beauty," in ch. 2 of my *Cosmos and Creator* (Edinburgh: Scottish Academy Press, 1981).

[6]References will be to his English translation, "Concerning the Face which Appears in the Orb of the Moon," in *Plutarch's Moralia, Volume XII,* Greek text with English translation and notes by A. Cherniss and W. C. Helmbold (London: William Heinemann, 1957).

[7]Darwin's comment on reading W. Ogle's translation of Aristotle's *On the Parts of Animals.* See F. Darwin, *The Life and Letters of Charles Darwin* (London: Murray, 1888), vol. III, p. 252.

[8]That this truth was pointedly and emphatically recalled to our generation by K. R. Popper (*Conjectures and Refutations* [New York: Harper and Row, 1968], p. 136) is somewhat strange as the validity of the notion of cosmos, the chief object of cosmology, is rather ambiguous in the Popperian perspective.

[9]*Aristarchus of Samos: The Ancient Copernicus* (Oxford: Clarendon Press, 1913).

[10]For text and translation, see ibid., pp. 351-413.

[11]See ibid., pp. 305-06.

[12]For texts and interpretation, see S. Sambursky, *The Physics of the Stoics* (New York: Macmillan, 1959), pp. 108-10.

[13]"Concerning the Face which Appears in the Orb of the Moon," in *Plutarch's Moralia,* vol. XII, p. 55.

[14]See *The Almagest,* Book 1, ch. 7, in *Great Books of the Western World* (Chicago: Encyclopedia Britannica, 1952), vol. XVI, pp. 10-12.

[15]Concerning these four ancient cultures, see chs. 1, 2, 4, and 5 in my *Science and Creation: From Eternal Cycles to an Oscillating Universe* (Edinburgh: Scottish Academic Press, 1974).

[16]The arbitrariness of this perspective, which has B. Farrington as one of its chief spokesmen in the West, has been convincingly set forth by L. Edelstein in *Roots of Scientific Thought: A Cultural Perspective,* edited by P. P. Wiener and A. Noland (New York: Basic Books, 1957), pp. 103-13.

[17]See A. C. Crombie (ed.), *Scientific Change* (New York: Basic Books, 1963), p. 885.

[18]W. Whewell, *History of the Inductive Sciences,* 3rd edition (reprinted; London: Frank Cass, 1967), vol. I, pp. 58-59.

[19]Published in 1913 and 1914, respectively (Paris: Hermann).

[20]Volumes 3, 4, and 5 were published in 1915, 1916, and 1917, respectively. The last five volumes saw print only in the 1950s. [Behind that almost forty-year delay there lies an academic scandal which is aired in my article, "Science and Censorship: Hélène Duhem and the Publication of the *Système du Monde,*" *The Intercollegiate Review* 21 (Winter 1985), pp. 41-49.]

[21]As reported in *Un Savant Français: Pierre Duhem* (Paris: Plon, 1936, p. 40), the still indispensable source on Duhem's life and work written by his daughter, Hélène Pierre-Duhem. [My indebtedness to it is duly registered in the Preface of my half-a-million-word-long monograph, *Uneasy Genius: The Life and Work of Pierre Duhem* (Dordrecht: Nijhoff, 1984).]

[22]Ibid., p. 48.

[23]Originally published in seven installments, between January 30 and April 30, 1903, in the *Revue générale des sciences.* As a book it appeared in the same year (Paris: A. Joanin) and again in 1905 (Paris: Hermann).

[24]A work of two large volumes (Paris: Hermann, 1905-06).

[25]Ibid., vol. I, p. 12.

[26]*La théorie physique. Son objet—sa structure* (Paris: Rivière). A second enlarged edition followed in 1914.

[27]To this richly documented, pioneering work, which has for its subtitle: *Essai sur la notion de théorie physique de Platon à Galilée* (Paris: Hermann), there are only three insignificant references in *Die Rettung der Phänomene: Ursprung und Geschichte eines antiken Forschungsprinzips* (Berlin: Walter de Gruyter, 1962), by J. Mittelstrass, a book which covers much the same topic and material.

[28]*Le système du monde,* vol. I, p. 130.

[29]Ibid., vol. I, p. 34.

[30]*La théorie physique,* a work which today is far more available in its English translation, *The Aim and Structure of Physical Theory,* by P. P. Wiener (Princeton: University Press, 1954; paperback reprint, New York: Atheneum, 1962). See pp. 24-27 of translation.

[31]*Les Origines de la statique,* vol. II, p. 290. Duhem, of course, does not advocate a *Deus ex machina.* A firm believer in evolution, though not in Darwinism, Duhem was fond of casting intellectual progress in evolutionary terms, and in this particular context he pointedly recalled Claude Bernard's portrayal of biological growth in terms of an *idée directrice.*

[32]See, for instance, *The Aim and Structure of Physical Theory,* p. 102. This

was a Pascalian attitude, quite natural on the part of Duhem who knew the *Pensées* by heart.

[33]See *Le système du monde,* vol. I, pp. 101-03.

[34]Ibid., p. 240.

[35]Theophrastus, Περι σημειων. Duhem also quotes Proclus concerning Theophrastus' admiration of Chaldean astrological lore. *Le système du monde,* vol. II, p. 274.

[36]"La théorie des marées et l'astrologie," *Le système du monde,* vol. II, pp. 266-390.

[37]In doing so, Duhem is most conscientious in acknowledging his debt to others who had published on the topic.

[38]In an unobtrusive display of his scholarship Duhem notes that the same idea is later voiced by Descartes. *Le système du monde,* vol. II, p. 274.

[39]["Planetary hypotheses"] in *Opera astronomica minora,* edited by J. L. Heiberg (Leipzig: B. G. Teubner, 1907), p. 71.

[40]*Le système du monde,* vol. II, p. 334.

[41]Ibid., vol. II, p. 366.

[42]"Concerning the Face which Appears in the Orb of the Moon," in *Plutarch's Moralia,* vol. XII, p. 59.

[43]*Le système du monde,* vol. II, p. 363.

[44]Ibid., vol. I, p. 27.

[45]"Concerning the Face which Appears in the Orb of the Moon," in *Plutarch's Moralia,* vol. XII, p. 213.

[46]*Le système du monde,* vol. II, p. 390.

[47]*Phaedo,* 97-101. Almost two thousand years later, Leibniz, in order to forestall the "mechanization of all" in terms of the new physics, suggested, with a reference to *Phaedo,* the working out of a mathematical physics in which the notions of purpose and good would be safeguarded. See *Leibniz Selections,* edited by P. P. Wiener (New York: Charles Scribner's Sons, 1951), pp. 69 and 320.

[48]*Le système du monde,* vol. I, pp. 11 and 43.

[49]This theme is developed in my Gifford Lectures, *The Road of Science and the Ways to God* (Chicago: University of Chicago Press, 1978; paperback edition, 1980), ch. 2, "A Lesson in Greek."

[50]"All of Aristotle's philosophy," Duhem writes, "claims that an absolute determinism reigns in the Universe. His conscience cried out that he was able to act, that it was in his power to produce or prevent certain results; therefore he had to admit that there was contingence down here. But through that illogical concession he broke the coherence of his doctrine. For if one wants to remain [within the Aristotelian context] consistent with oneself, one either has to reject the astrological axiom which contains within itself all the Aristotelian physics and metaphysics, or one has to hand the world over to an absolute fatalism." *Le système du monde,* vol. II, p. 297.

12

Christian Culture and
Duhem's Work

In 1953, in an often-quoted lecture given at Harvard, Herbert Butterfield warned about a momentous shift in Western education. Traditional study of Greek and Latin authors was being rapidly replaced by the study of the classics of science as the framework in which to examine basic questions about human existence. The vanishing of courses in Latin and Greek has since become a doubtful distinction of ecclesiastical training as well. The great sages of ancient Greece and Rome are, of course, still presented to college students, but only in surveys of Western civilization which invariably end in encomiums of the relativization of values.

Plato for one had full awareness of the fact that the quest for

Reproduced with permission from *The Dawson Newsletter* 3 (Summer 1984), pp. 6-8, where it appeared under the title, "An Author's Reflections," in connection with the publication of his work, *Uneasy Genius: The Life and Work of Pierre Duhem* (Dordrecht, London and Boston: Martinus Nijhoff Publishers, 1984), pp. xi + 470, illustrations, notes, list of Duhem's publications, name and subject index.

commitment to absolute ethical values and ontological truth, which was the chief aim of the Socratic movement, had as its only real alternative the relativizing of all values. He also spoke in startlingly modern terms of the proliferation of symptoms of wilfulness once respect for absolutely valid norms and truths becomes unfashionable. Plato could not be explicit in describing that runaway relativization as an indulging in ever new patterns for novelty's sake. He could not do it because he lived 2,000 years before the rise of that kind of modern science whose sole aim is the study and discovery of patterns. It would not, however, have required the acumen of a Plato to perceive that exclusive preoccupation with patterns, which are always configurational and therefore quantitative or numerical, can but stifle interest in the notion and reality of that very purpose which is inherent in values. With the banishing of a systematic inquiry about purpose, religion too is bound to be banished—a conclusion which is as old as Plato and as contemporary as space travel.

Largely because of Socrates' heroic stance on behalf of purpose as expressive of the human soul, classical antiquity was essentially religious. It was therefore a fertile ground for Christianity even though the latter assigned to the human soul individual immortality and irrevocable once-and-for-all destiny—tenets wholly unacceptable to classical paganism. If our times have been called post-Christian, it is largely because of the overwhelming preoccupation, under the impact of science, with mere patterns. The reason for this is a worshipful appraisal in which science becomes sheer scientism, that is, the proposition that only scientifically measurable parameters are worthy of consideration. Such is the real issue in that often celebrated warfare between science and Christianity, and especially dogmatic Christianity. While this is largely overlooked by Christian divines watching the world through glasses of naïve optimism, quite the opposite was true of those who about 200 years ago laid the groundwork for that post-Christian era of ours. A Condorcet and later a Comte, to mention only two names among many, were fully aware both of the logic that ties Christianity to abiding purpose and of the ease with which an ever more successful science might be exploited on behalf of the exclusive validity of mere patterns, a tenet lying at the very heart of scientism.

Thinking of themselves as ones riding the wave of the future tied to science, Condorcet and his cohorts found it most natural and logical to charge the Christian past with enmity to science as its gravest crime against human progress. Thus while a Kepler and even a Leibniz still had some notion of medieval men of science, a hundred years later it became fashionable to speak of the rise of science as an early seventeenth-century secular miracle, a sort of *Deus ex machina* in which the machine readily took the place of God. The fashion had become so pervasive that even such a heroic modern defender of purpose as Bergson blithely spoke in his *Evolution créatrice* of the coming down of science from heaven on the inclined plane of Galileo. Secularization had found a ready equivalent even for Jacob's ladder. Implied in that simile was that darkness was the true characteristic of the Christian centuries, that is, of the Middle Ages, preceding that great secular miracle.

Today it no longer passes for unquestionable scholarship to speak of the darkness of the Middle Ages. Once the rationalizing madness of the Enlightenment had run its feverish course, the artistic value of Gothic cathedrals was discovered by the Romantic movement and the Comteans themselves began to praise the social coherence which the Middle Ages owed to the Catholic Church. Half a century later the great scholastic philosophers entered, one by one, into the respectable nomenclature of secular scholarship. There is, however, one area, science, where the alleged darkness of the Middle Ages is still a fairly respectable myth. Challenges to that myth are not necessarily dismissed with condescending smiles but they are hardly ever given serious consideration. Such is a most telling fact in an intellectual context in which massive documentation commands the highest accolades.

Nothing short of monumental was the documentation which Pierre Duhem (1861-1916) provided on behalf of the indispensable role which the medieval Christian cultural matrix played in the rise of modern science. Yet, although in 1961, the hundredth anniversary of Duhem's birth, a blue-ribbon gathering of historians of science heard Henri Guerlac of Cornell University declare that "Pierre Duhem is the teacher of us all," not a single symposium was held in Duhem's honour. Rather, all over the world Ernst Mach was celebrated in 1966, which was the fiftieth anniversary not only of his death but also of Duhem's death! Nothing was said in that year about the fact that Mach, a self-declared antagonist of Church and Christianity, refused to the very end to use the word "medieval" in the few references in which he grudgingly acknowl-

edged Duhem's researches. Whatever there was scientifically valuable in the fourteenth century, it remained for Mach a credit to the
spirit of the Renaissance.

It tells much of the entrenchment in the academia of positivism
(logical and other) which Mach had established among philosophers
and historians of science that Duhem has, as a rule, been recognized only as an ally of Mach. Yet Duhem most explicitly endorsed
Aristotelian ontology as the sole ground on which the purely
positivist procedures of mathematical physics can be tied to reality.
A similar misinformation gained ground about Duhem as the historian of science. Although he was a staunch defender of the continuity of intellectual development, he has recently been presented
as a forerunner of those paradigmists who fragment the unity of
science into the succession of mutually incommensurable perceptions which overthrow one another as do revolutions.

Even an outsider might suspect some tendentiousness in the fact
that during the 1950s, 1960s, and 1970s, which saw the publication
of countless monographs on major and minor historians and philosophers of science, no systematic study appeared on the life and
work of Duhem. Of course, the task would not have been easy. In
Duhem's lifework three fields—theoretical physics, philosophy of
physics, and history of physics—are intimately interwoven. Moreover, the amount of his printed work stretches over 25,000 pages, a
forbiddingly vast quantity to be digested in one volume which must
accommodate also something of his vast correspondence, and an
enormous background material. Yet in that apparent disinterest in
Duhem, there were at play factors that had nothing to do with the
inherent difficulty of the task in question. Of those factors a good
glimpse may be gathered from a quick review of his life.

Like many of his compatriots brooding over the defeat of 1870,
young Duhem saw intellectual excellence as a patriotic duty. His
patriotism was also marked with a deep attachment to the Catholic
faith which he imbibed at home and at Collège Stanislas (Paris),
where he received his secondary education. Secularism and anti-
clericalism were rapidly gaining ground by the time Duhem in 1882
entered the Ecole Normale, the citadel then of French higher
education. He chose that school among the various *grandes écoles*
as the one which seemed to him most germane to his ambition of
becoming a theoretical physicist. Duhem most explicitly saw his
commitment to theoretical physics (and in particular to its then

latest branch, thermodynamics) as a patriotic service at a time when French physics had shrunk to mere experimentation, and the great line of French theoretical physicists from Laplace to Fresnel had no worthy continuators.

Duhem's commitment was matched with extraordinary intellectual excellence and also with excessive, almost naïve idealism. He thought that all leading men of science were disinterested servants of truth alone. He was in for a rude shock. Already in his second year at the Ecole Normale he had formulated what later became known as the Gibbs-Duhem equation, a cornerstone of physical chemistry. Unfortunately for him, the equation amounted to a rebuttal of the maximum work principle which Marcelin Berthelot had borrowed from the Danish chemist, Julius Thomsen, and had imposed as his own brainchild on much of the French scientific establishment. In his idealism young Duhem did not reckon with the fact that by 1885 Berthelot had for years been the overlord of French higher education and the chief disburser of the best academic posts and emoluments. Duhem's doctoral dissertation, which is now in the "Landmarks of Science" series, was rejected by the Sorbonne where Berthelot's wishes were so many commands. Moreover, it was also decided on the highest level that no chair in Paris would ever be given to Duhem, although he had successfully defended two years later another brilliant thesis before a jury of pure mathematicians who were unafraid of Berthelot and eager to remedy a rank injustice.

While unjust treatment and "violation of human rights" suffered by scientists have for some time been a favorite topic for historians of science, Duhem's case was never investigated and partly for a reason that should be obvious. Berthelot belonged to a group (the heavy representation of members of the Grand Orient in the Mitterrand government is in direct succession to that group) which is still powerful enough to dissuade scholars from airing certain dirty linen. The same group could only be filled with horror when, from 1904 onwards, Duhem began to unfold startling evidences on behalf of the medieval Christian matrix of the origin of modern science, that is, of Newtonian mechanics. Duhem did that heroic work against enormous odds. He was banished to provincial universities, first to Lille, then to Rennes (then the most backwater university in France) and from there to Bordeaux. He spent his last twenty-two years there, giving his advanced courses time and again to almost empty classrooms. In addition, even in Bordeaux, one of the three best provincial universities then in France, he was deprived of adequate library facilities. Also, once he caught sight of those medieval scientists, he had to teach himself medieval paleography (with its many variants) and copy thousands of pages from hundreds of manuscripts which the big Parisian libraries sent him, at first with some reluctance. In this age of xerox machines, word processors, and microfilms, it is difficult to imagine the magnitude of Duhem's labors which issued in a dozen thick volumes on the history of medieval statics, dynamics, and related sciences. Yet, while his researches greatly enhanced the glory of the fourteenth-century Sorbonne, the Sorbonne preferred to take no notice. No wonder. In a France ruled intellectually by Renan and Co. it was a *lèse majesté* to point out that Buridan and Oresme formulated the impetus theory—an anticipation of Newton's first law and the basis of all modern physics—precisely under the impact of their belief in the Christian dogma of creation out of nothing and in time. If the glory redounding to the Sorbonne was also a Christian glory, it was not to shine in the Sorbonne of the twentieth century.

Had Duhem not had close friends among the older generation of French mathematicians and a growing renown outside France, he would have never been given a signal compensation, namely, his election in 1913 as one of the first six non-resident members of *Académie des Sciences.* The most influential among those friends, Darboux, perpetual secretary of the *Académie,* died, however, soon after Duhem's sudden death in 1916. The result was a discontinuation of the publication of Duhem's immortal opus, *Le système du monde,* on the history of physical and cosmological theories from Plato to Copernicus.

Efforts to have the last five of the ten volumes of the *Système* published in the 1930s were thwarted by factors in which the anti-Christian government policies of the Front Populaire and the scientists supporting them were unmistakable.[1] The publication could take place only with the advent of a different atmosphere in Paris in the 1950s. Although posthumous by forty years, those last five volumes were still an extraordinary storehouse of fresh information, a fact acknowledged even by those, such as A. Koyré, who did their best to undercut the reliability of Duhem's main conclusions.

The principal aim of Koyré's efforts was to save the eminent place customarily accorded to the Renaissance in modern historiography. The Renaissance as a vote against Christianity seemed to lose its weight when deprived of scientific content. Hence the efforts of Koyré to rehabilitate scientifically the Renaissance, and the efforts of all those historians and philosophers of science who found his lead "liberating." The price of those efforts to cast suspicion, directly or indirectly, on the value of Duhem's work turned out to be too high even for some rationalists and positivists. For Koyré could make his case plausible only in terms of a Darwinian re-casting of intellectual history in which ideas are haphazard mutations and locked in a senseless life-to-death struggle for survival in purely pragmatic terms with no intrinsic truth content. The secular equivalent of Jacob's ladder did not rise much above the lowest ground.

It is against this background that the drama of Duhem's superhuman intellectual struggles and researches should be seen. That he was neglected even when, during these last few decades, Catholic historians and philosophers of science began to be fairly numerous, is a sad sign of the times. My call for the formation of a Duhem Society made in 1978 before a large audience on a prominent American Catholic campus fell upon deaf ears. Apparently, the lure of secular applause all too often blocks sounder perspectives and can deaden consciences faced with unpopular objectives and duties. Among the latter is an unflinching devotion to the great cultural contribution of Catholicism. Thanks to Duhem, the evidence on behalf of the Christian matrix of the birth of science is unmistakable. It is only on that Duhemian basis that it became possible to argue in a historical perspective, as was done in my Gifford Lectures, that creative science always presupposes, implicitly or explicitly, a world view which is anchored in the Christian dogma of creation. Science is possible only if the universe is both rational, that is, intelligible, and is also created, that is, contingent, and therefore to be understood through empirical investigations that exclude an *a priori* approach. The converse of this thesis is that whenever the epistemology of a rebuttal of the cosmological argument is grafted on to the scientific method, the result is a potential fiasco for scientific progress. This is evident in the legislation given for science by a Hume, a Mill, a Mach, and in our times by the logical positivists, to say nothing of the paradigmists and other conceptual Darwinists.

Catholic intellectuals, still to a large extent perplexed about the true position and message of science, may do well to meditate on the life and work of Duhem. So far only a very few have done so. One of them was none other than Christopher Dawson. That for the past twenty years his work has become a practically unknown entity on Catholic campuses is part of a story of which the Duhem-story, covering both his life and the subsequent reception of his work, is a major and poignant chapter. Without attention to the Duhem-story there will be little hope for a reasonable Catholic answer to the threatening dichotomy of two cultures. The resolution of that dichotomy in a third culture, anchored in broad cultural considerations, calls, in this age of science, for an in-depth view of what science is truly about, structurally and historically. Perennially valid answers have been provided by Duhem to that problem. Hence the instinctive appreciation of Duhem's work on Dawson's part. Would that this perceptiveness might find many imitators in these times of desperate struggle to keep Catholic culture, past and present, visible. Only then will that culture be noticed by the up-coming Catholic generation which thirsts for more refreshing waters than the ones now being marketed under pirated versions of the sacred label of Vatican Council II.

[1]For documentary evidences, see my article, "Science and Censorship: Hélène Duhem and the Publication of the *Système du monde*," *Intercollegiate Review* 21 (Winter 1985), pp. 41-49.

13

On Whose Side Is History?

To ask, in these times, the question "On whose side is history?" is to risk sliding into politics—a discussion of capitalism versus Communism. Nikita Khrushchev's warning to Americans ("We will bury you") was a particularly crude answer to this question in political terms, but it was also a statement of the fundamental Marxist dogma according to which history is merely the offspring of the tools of production. History—that is, final victory—will therefore be on the side of those possessing the greatest competency in forging the tools best suited to produce the best society.

Unlike the concept of the best society, the making of the best tools is hardly a debatable matter. Tool-making is no longer a matter for tinkerers, but has become the task of the most exact sciences. Now Marxism has always claimed to be a scientific ideology —indeed, the only true scientific ideology. In the preface to the second edition of *Das Kapital* Marx claimed to have offered a theory of economics no less exact than Newton's celestial mechanics. Countless Marxists followed suit in repeating the same claim

Reprinted with permission from *National Review*, August 23, 1985, pp. 41-47.

from different perspectives: The title of this essay is merely a variation on the title of the book *History Is on Our Side,* published about forty years ago by a prominent British Marxist, Joseph Needham.

That Professor Needham was, and still is, a Marxist is not my chief reason for recalling him and his book; rather, it is that just when Professor Needham—a biochemist by training—published that book, he was embarking on a monumental project on the history of science in ancient China. As a Marxist he felt certain that the long subjection of the Chinese people to feudal lords and a feudal type of economy was the reason why science had not been born in China. But by the time Professor Needham finished (around 1955) the second volume of his now nine-volume work, he had reached a totally different conclusion. As he put it in the final section of that second volume, the Chinese of old failed in science because they had failed in theology. Having rejected, sometime in the early second millennium B.C., their belief in a personal, rational, and transcendental Creator, a Lawgiver, the Chinese lost confidence in the ability of the human mind to fathom the laws of nature.

So much for Professor Needham and so much by way of advance apology if I happen to come near politics, as I try to provide, in my capacity as a historian of science, an answer to the question "On whose side is history?"

We might begin the inquiry by giving history itself a brief look. The origins of historiography, like the origins of many other branches of learning, go back to the Greeks—Herodotus, Xenophon, and Polybius. They were excellent narrators, consummate analysts, and even psychoanalysts and psychohistorians. Yet none of them raised the question of why human history exists and whether it has a meaning. On the sole occasion on which Polybius came fairly close to raising that question, he answered it in such a way as to make of it sheer mockery.

For modern Western man the question "On whose side is history?" may invite a definite and meaningful answer. Not so for Polybius. In relating the Roman conquest of Greece, he put history on the side of the Romans—in his eyes, a nation of barbarians—but only for a while: The Romans, Polybius wrote, would ultimately go down in defeat because nothing—no power, no society, no army—is above the basic law of history. That law was for Polybius and the

Greeks of old the law of the wheel, or the inevitable sequence in which birth and death, progress and decay, success and failure alternated to no end.

The symbol of that chain of ups and downs was the *gammadion,* the Greek word for swastika, a standard decoration in many Roman and Greek mosaics. In Sanskrit *swastika* is a compound word of *su* and *asti,* that is, of *well* and *being,* or *well-being.* Whether this being caught in the endless treadmill of ups and downs, of endless births, deaths, rebirths, and deaths again, can logically be called well-being is not the purpose of this essay to decide. Tellingly, the sign was most popular among both the Greeks and the Romans when their fortunes seemed to be on the rise, when they felt themselves riding the crest of the wave.

In adopting the swastika as their emblem, the Nazis were certainly logical. Their neopaganism harked back to pre-Christian paganism. The Nazis were logical also in stating that the New Europe they aimed at establishing would not last forever. Just as the swastika was turning, their fortunes too would eventually take a downward turn. Luckily, and contrary to their expectations, their downfall occurred not after a thousand years but rather in a mere fraction of that time.

But to return to the Greeks and Romans. One of the great military contests between them took place in Magna Graecia, or Sicily; at Syracuse, to be specific. On account of its geographical location alone, Syracuse was a hard nut to crack; it had to appear downright impregnable when its natural defenses had been supplemented by the efforts of Archimedes, the greatest scientist-engineer of all antiquity. His catapults hurled enormous rocks. His huge cranes picked up Roman ships from the sea as soon as they reached the walls of Syracuse. Yet none of these engineering triumphs could persuade the Romans to lift the siege of Syracuse. The case might have been quite different if Archimedes had been an expert on ballistics. On this point, the study of motion, which is the very soul and foundation of physics, the Greeks made no advances whatever.

Nor, for that matter, did the Chinese. The case of the Chinese is all the more tantalizing because they had to their credit three inventions—block-printing, gunpowder, and magnets—that Francis Bacon naïvely claimed for his own times and in which he saw the beginning of science. It would indeed be tempting to picture the ancient Chinese sailing along the Aleutian Islands, carrying small cannons, as the Spanish conquistadors were to do, and colonizing Alaska, British Columbia, California, the Rockies, the Great Plains, the Ohio Valley, and New England for good measure. North American atlases would now list New Shanghai instead of New York, New Peking instead of New Jersey, and *horribile dictu,* New Mongolia instead of New Haven. As for Boston, no more than a hyphen (Bos-Ton) might have been needed. All this could easily have happened if the Chinese of old had been scientific in the sense of steadily developing their inventions. History—global history—would have been on their side.

History first became global history in the seventeenth century, when Western history became global history. But Western man could not have conquered the globe had he not had science and technology on his side. (Whether the simultaneous growth of dynamics and the military science of ballistics has been an unmixed blessing is another question.) At any rate, history for the past three hundred years has been on the side of the West, and remains so. Furthermore, serious challenges to the global domination of the West are made essentially on a scientific basis. Without Soviet scientists there would be no Soviet superpower, and without the wide availability of modern technology there would be no wars of "liberation" around the globe. Even China with her enormous superiority in manpower will not represent a tangible threat to the West until Chinese science and technology catch up with Western science and technology, a feat not easily accomplished given the rate at which progress in these areas is occurring in the West.

How did the West acquire in the first place its astonishing scientific lead? It did so by rejecting what was distinctly and fundamentally pagan in ancient Greek science. This may sound a strange proposition to anyone who thinks that science is no more than a manipulation of quantities which in themselves are neither pagan nor Christian. But science—especially the science of motion—has always implied, and will forever imply, certain basic philosophical presuppositions.

Motion for the Greeks was eternal and uncreated; such was certainly the belief of Aristotle, the foremost theorist of motion in Greek antiquity. When Aristotle became fully known in the West in the late thirteenth century, Western Europe was deeply Christian. The newly established universities taught everything that could be known (that is why they were called universities) but they taught it from the Christian perspective, the cardinal and essential point of which is the dogma of creation in time. This dogma means that the past history of the world is finite, that there is a point back in history where all motion had to start. This perspective was inconceivable, or at least repugnant, to the Greeks, but was most natural for Christians. This same idea—that everything, all motion, had to have an absolute starting point—was also the natural start of science as we know it: a science in which one discovery generates another discovery, a science that cannot be stopped.

In intellectual history, actual starting points are difficult to specify. Who was the first novelist? Who was the first economist? Who was the first military strategist? Who was the first impressionist? Unlike these questions, which hardly admit an unambiguous answer, it is fairly easy to identify the first modern physicist. He was Jean Buridan, professor of philosophy at the Sorbonne around 1330. His lectures were, as were all university lectures at that time, commentaries on the works of Aristotle. In his consideration of the passage where Aristotle wrote, in his cosmology *On the Heavens,* that the universe and the motion of the stars were eternal, Buridan did two things. First, he voiced his profound disagreement with Aristotle. The notion of the eternity of the universe and all motion (including the motion of the stars) could not be true because it contradicted Christian Revelation, according to which God created the universe in time. Second, Buridan offered a splendid speculation about the origin of the motion of the stars:

> When God created the world, He moved each of the celestial orbs as he pleased, and in moving them He impressed in them impetuses which moved them without His having to move them any more except by the method of general influence whereby He concurs as a co-agent in all things which take place; . . . these impetuses which He impressed in the celestial bodies were not decreased nor corrupted afterward, because there was no resistance which would be corruptive or repressive of that impetus.

This statement of Buridan, the equivalent of Newton's first law of motion, is—conceptually but also historically—the beginning of modern science. Buridan's statement was copied in countless manuscripts, which students at the University of Paris carried far and wide in Europe; it was repeated and commented upon many times during the fifteenth and sixteenth centuries. This statement, or its equivalent, was well known to Galileo and Descartes, who are still credited with the formulation of Newton's first law by historians of science who have failed to do their homework properly.

That is a glaring miscarriage of justice, but only one of the many perpetrated in the academic world, which has also refused to take proper notice of one of the most important discoveries in the historiography of science. That discovery took place in Bordeaux in early 1906 in the modest home of a professor of physics at the University of Bordeaux. Pierre Duhem, then 45 years old and a widower, already had a world reputation as an expert on thermodynamics and continuum mechanics. Until two years before, Duhem had taken for granted that there was no "science" in the Middle Ages; that the theological mentality of medieval universities was inimical to creative scientific thought. He had taken for granted the notion that the Middle Ages had been a period of scientific eclipse because, in his time, this notion had been for two hundred years a hallowed tenet of cultural respectability. It derived partly from the Reformers' scorn for medieval Catholicism and partly from the hostility of the leaders of the French Enlightenment to anything Christian. While by 1900 the Middle Ages had been rehabilitated from the viewpoint of the arts, of social organization, of welfare, and of literature, nobody dared to think that there was any science, let alone good science, to look for in the Middle Ages.

Pierre Duhem, the theoretical physicist, was led to the science of the Middle Ages because he felt that theoretical physics, like anything else, is fully and properly understood only if taken not just in itself but also in its historical development. In tracing the concept of virtual velocity—the foundation of dynamics—farther and farther back in history, Duhem had to go beyond Galileo to Galileo's teachers, Benedetti and Stevin, and beyond these to Cardan. The latter's cryptic reference to a mysterious Jordanus helped Duhem discover an until then unsuspected period of science culminating in the work of two fourteenth-century professors of the Sorbonne, Buridan and his disciple, Nicole Oresme, who subsequently became bishop of Lisieux.

Duhem's was not an easy and straightforward search. He had no previous scholar to lead him. He had no printed material to rely upon. He had to go through countless medieval manuscripts that nobody had touched for centuries. He had to teach himself medieval handwriting, or rather the many forms of medieval Latin stenography, which varied from area to area, from century to century. He had no secretaries, no research assistants, no eager graduate students, no interlibrary service, not even ballpoint pens. At every half-line he had to dip his pen into the inkwell, adjust the measure of ink taken up by the pen, shift the blotter, and resume writing. He had no photo-copying machines, no dictaphones. He filled 120 notebooks, of two hundred pages each, with long excerpts from medieval manuscripts. He did all that with an often-trembling right hand which he had to steady with his left so that his handwriting might be decipherable not only by himself but also by the printer.

The effort was monumental and so was the result; the ten huge volumes of *Le système du monde*, the three huge volumes of his Leonardo studies (reprinted last year for the third time), and the two huge volumes of *Les origines de la statique*. As one historian of science said of Duhem: He revolutionized the history of science by creating the historiography of science.

Much of this may produce a touch of disbelief. How is it that, if such a great feat had been accomplished almost eighty years ago, nothing has been said of it in typical college surveys on the history of science, or in general surveys of early modern Europe? The answer to this question is simple and was in fact given as early as 1911, in a sumptuous volume on the history of the Sorbonne written by Louis Liard, director of French higher education. Liard was enough of a French chauvinist to present the Sorbonne in as much glory as possible—but only so long as that glory was not a Christian glory. So he kept silent on the subject of Duhem and his epoch-making discovery, although he was fully aware of Duhem's major publications.

Duhem himself did his best to alert his fellow Catholics, especially the teaching members of the clergy, to what was at stake. He urged the establishment of two chairs at the Institut Catholique, one for the philosophy of science and the other for the history of science. Familiar with the chief strategy of the antagonists of Christianity, who had for a century been expropriating science for their own purposes, Duhem saw that a pivotal task of Christendom consisted in setting forth the true cultural history of the West even if that history would be branded as revisionist history in many circles.

In that true history, science would no longer appear as something that could have arisen only when Christianity had been discredited; rather, science would be recognized as owing its very birth to Christianity. In other words, Duhem called for exposing a chief and carefully cultivated trickery of the Western intellectual establishment: the "scholarly" lie that science and Christianity are irreconcilable. By 1936—twenty years after Duhem's death—a concerted effort was under way among influential historians of science to discredit Duhem's scholarship either by silence or by condescending phrases; and that effort is under way still. It will never be forgiven him in some circles that like a David he slew a Goliath—a lie of gigantic proportions according to which Christianity had to be discredited so that science might arise.

This story—which is fully exposed in my recently published book, *Uneasy Genius: The Life and Work of Pierre Duhem*—is not mentioned here as a reminder of old accounts to be settled: Such would not be a Duhemian attitude at all. History, or the history of science, had for Duhem three main lessons. One was a rather sad lesson. Sad indeed is the history of science insofar as it shows—to recall a phrase of Duhem—the repeated apostasies of the human mind, the repeated yielding to pleasing but obvious fallacies taken for basic frameworks of explanation. Such a fallacy was at work when Newtonian physics was taken for the proposition that everything is machine; another is at work when relativistic physics is taken for the claim that everything is relative; still another is at work when quantum mechanics is taken for a denial of causality.

Another lesson, a potentially encouraging one, is related to the future. The history of science was, in Duhem's eyes, a process guided by an *idée directrice,* which he as a staunch Catholic was not ashamed to call simply Divine Providence.

But there was a third lesson as well, which dawned on Duhem in 1915 when the enormous bloodletting of World War I led him to perceive science as a sin against the Holy Spirit. By then the number of casualties on both sides was well over three million, a slaughter made possible only by the contribution of science to military technology. Duhem did not live to write the lecture series he had planned on man's sinful misuse of science, nor of course did he live to see the 25 to 30 million dead of World War II.

It should not be difficult to guess what he would say today. As a man of principle and a sworn enemy of any appeasement with half-truths—let alone with plain lies—he would not view ruthless dictatorships as being merely aged bureaucracies. He would also have a word or two for Western democracies. Although a fierce defender of academic freedom and of professional independence from imposed constraint, religious or political, Duhem was no friend of democratic politics. He had a high regard for the idea of democracy, but he knew that modern democracies are often exercises in pragmatism in which success is confused with truth. It could not have escaped him, a sharp logician, that pragmatism is frequently a mere bargaining for time.

That is why he would say today that the future, or history, is not necessarily on the side of the West, in spite of the West's unquestionable superiority, technological and scientific, over the rest of the world. Pragmatism may be unsurpassable in mustering know-how, for this is what pragmatism is about. But pragmatism is know-how without inspiration, unless it is the inspiration of mere profit. It is without a soul and, for all its determination, it is never sufficiently determined. It may provide new sets of superweapons, but it is unable to produce judicious and principled persons who will not abuse their potentially destructive science. Pragmatism is the great breeding place of professional appeasers, who will blink first when the great confrontation presents itself.

For, ultimately, the question "On whose side is history?" will not be determined by science, or rather by proficiency in science. History will be on the side of those who have the greatest determination, the most steadfast spirit—moral characteristics that are never the pride of places and cultures in which the highest premium is put on immediate self-gratification.

But is science merely a passive tool in the great historical drama conjured up by the question "On whose side is history?" Is science going to be but a tool that creates security for the affluent, or makes life easy for the tyrant, or turns all men into dehumanized robots? Is there in science something that would play an active role, and a role for the better, as mankind heads into an increasingly scientific future, from which there is no turning back?

A glance at the present world situation may suggest that there is indeed an active and constructive role for science. In these days, when all eyes are riveted on the Middle East, nothing is so tempting as to look upon Muslim revivalism—which is boiling from Indonesia to Morocco, from Afghanistan to Nigeria—as an upsurge of transient fanaticism that will blow itself out in the not-too-distant future: a perfectly pragmatic appraisal, representing pragmatic superficiality in its most highly developed form.

What is happening in the Muslim world is not so much an outburst of fanaticism as a frantic last-ditch effort to ward off the specter of—well, not of capitalism, not of Communism, not of hedonism—but of science. What is occurring in the Muslim world today is a confrontation, not between God and the devil, identified with capitalism or Communism, but between a very specific God and science which is a very specific antagonist of that God: the Allah of the Koran, in whom the will wholly dominates the intellect. A thousand years ago the great Muslim mystics al-Ashari and al-Ghazzali denounced natural laws, the very objectives of science, as a blasphemous constraint upon the free will of Allah. Today, the impossibility of making ends meet without science forces the Muslim world to reconsider its notion of Allah. It is an agonizing process, which, in spite of the bloodshed, may, in the long run, bring a more rational mentality to troubled parts of the world.

In speaking about the active role of science in shaping history, the status of science in the Soviet Union provides another telling insight. It is little remembered today that in 1952, when Stalinist ideological control was at its peak, physicists and physicists alone were able to maintain some independence of thought. The Marxist state, which rode roughshod over any and all opposition, had to yield when told by leading physicists that they could deliver to Party and State the needed technological breakthroughs only if they were allowed to think freely.

China, the no longer sleeping giant of the future, provides another insight into the constructive role science can play in history. The rupture between Communist China and Communist Russia represents more than regional rivalry and ethnic antagonism. Communist China had to open up to the West, had to liberalize its daily life, had to acknowledge, however modestly, the role of the individual, precisely because it needed the science that only the West could deliver and that can only be handed effectively through a greater infusion of the Western spirit.

Meanwhile, in the West, the active role of science is all too obvious. The West is in the throes of the abuse of science on a vast scale. The threat of cancer-producing chemicals has been daily news for the past dozen years. The threat of acid rain, of dioxide, of aerosol spray is too frequent a topic to need more than mere mention. While we humans may have saved ourselves from being poisoned by exhaust fumes, some of our finest forests are perishing. No less a threat than the prospect of nuclear holocaust is the threat of possible misuses of genetic engineering. Perhaps the ever present possibility of ever greater misuses of the ever finer tools provided by science will spark a renewed interest in old-fashioned moral fiber. The Three Mile Island debacle demonstrates the point: The trouble there was not so much with nuclear engineering as with the functionaries who, for the sake of profit, left incompetent technicians in key posts of control.

Science may spark a moral reawakening here in this very pragmatic part of the world. Science may inspire greater rationality in lands where blind will is worshipped. Science may extort more and more freedom from institutionalized oppression. Science may, for the sheer necessity of survival, secure greater respect for thousands of millions of individuals who have been treated in recent decades like so many ants. Science may banish witchcraft from its last strongholds. Science, provided it is not abused, may provide hope for a better future, which will see history coming down not on this side or that side but on all sides—on the side of mankind as a whole.

The likelihood of this eventuality rests largely with the scientists themselves. They often boast that, whereas artists and philosophers—to say nothing of politicians—usually wind up in mutual incomprehension at international meetings, men of science can readily understand one another. These scientists, however, rarely remind themselves that their language—the language of quantities —is a very limited language; and much less do they trouble themselves to recall that the origin of that language is in a broadly shared vision of the whole material world as the embodiment of quantities.

This vision first became a commonplace in those same Middle Ages that gave the world a Buridan and an Oresme, when—to quote E. Curtius, a Protestant and one of the foremost students of medieval literature—the most often quoted phrase of the Scriptures was not a phrase from the Gospels or from St. Paul but one from the Book of Wisdom. There, God is praised for having disposed everything according to measure, number, and weight. Of course, the medievals knew that God was not restricted to any specific system of numbers, weights, and measures. They knew that the actual universe, as a creation of God, could have been otherwise, that it could not be necessary, but had to be contingent or conditioned on a divine decision.

That the world—the universe—might have been different is one of the great lessons of twentieth-century science. No wonder that Einstein, around 1950, felt the need to reassure a friend that in spite of his science he had not yet fallen into the hands of the priests. He also admitted that his science was lending itself to an art in which Catholic priests had long excelled: the art of natural theology, of constructing rational proofs of the existence of God, an art long since given up by liberal Christianity and recently abandoned by neo-modernist Catholicism as well.

Does all this mean that history will be on the side of dogmatic Christianity, and partly because of science? The opportunity is tremendous. To make the most of it, a concerted effort is required on the part of Christian scholars still holding fast to dogmas that have always been matched by good natural theology. If they fail to make that effort they will become easy prey to powers far less organized than the one once represented by Khrushchev as being ready and willing to bury them. They will hang separately if they will not hang together, and if they do not hang on with all their strength to the consistent defense, be it the rock called Peter, of the dogma of the createdness of all.

That rock may have had its momentary quarrel with individual scientists, or rather with men of science who turned their good science into bad philosophy and theology. That rock will pity men of science who, in Toynbee's words, are "trying to catch an angel in their butterfly net." That rock will safely outlive self-styled Christian men of science who, with Alistair Hardy, a Nobel laureate and Templeton Prize winner, "do not think there can be any future for orthodox Christian beliefs." At any rate, that rock stands firmly for the world view which provided the indispensable spark for the birth of science and with which science must side if it is to remain true to its ongoing creativity. Only when steeped in that world view can scientists marshal the moral strength commensurate with their opportunity to let history be on everyone's side.

Index of Names

Note on the Author

Stanley L. Jaki, a Hungarian-born Catholic priest of the Benedictine order, is Distinguished Professor at Seton Hall University, South Orange, New Jersey. With doctorates in theology and physics, he has for the past twenty-five years specialized in the history and philosophy of science. The author of sixteen books and over seventy articles, he served as Gifford Lecturer at the University of Edinburgh and as Fremantle Lecturer at Balliol College, Oxford. He is recipient of the Lecomte du Nouy Prize and has lectured at major universities in the United States, Europe, and Australia. He has recently been elected *membre correspondant* of the Académie Nationale des Sciences, Belles-Lettres et Arts of Bordeaux.